Jefferson's Shadow

Jefferson's Shadow

The Story of His Science

Keith Thomson

Yale
UNIVERSITY PRESS
New Haven & London

Published with assistance from the Annie Burr Lewis Fund.

Yale University Press books may be purchased in quantity for educational, business, or promotional use. For information, please e-mail sales.press@yale.edu (U.S. office) or sales@yaleup.co.uk (U.K. office).

Set in Bulmer type by Keystone Typesetting, Inc.
Printed in the United States of America.

Library of Congress Cataloging-in-Publication Data

Thomson, Keith Stewart.
Jefferson's shadow : the story of his science / Keith Thomson.
p. cm.
Includes bibliographical references and index.
ISBN 978-0-300-18403-7 (cloth : alk. paper)
1. Jefferson, Thomas, 1743–1826—Knowledge—Science. 2. Jefferson, Thomas, 1743–1826—Knowledge and learning. 3. Science—History—18th century. 4. Science—History—19th century. I. Title.
E332.2.T56 2012
973.4′6092—dc23
[B]
2012022590

A catalogue record for this book is available from the British Library.

This paper meets the requirements of ANSI/NISO Z39.48–1992 (Permanence of Paper).

10 9 8 7 6 5 4 3 2 1

For Linda

Contents

Preface and Acknowledgments

My fascination with Thomas Jefferson began when, as a graduate student newly arrived in the United States from England, I read for pleasure Adrienne Koch and William Peden's *The Life and Selected Writings of Jefferson* in the Modern Library series. My interest was rekindled when, as president of the Academy of Natural Sciences of Philadelphia, I became the nominal custodian of Jefferson's fossil collection and attended his 250th birthday celebration at Monticello in 1993. Ten years later, when Andrew O'Shaughnessy nominated me to become a visiting fellow at the International Center for Jefferson Studies at Monticello and to write a book on Jefferson and natural history, I jumped at the chance. (The book, *A Passion for Nature,* was published in 2008.) In the wonderfully supportive atmosphere of the center, it seemed only natural to fill a gap in the otherwise extensive literature on Jefferson by expanding that work into a volume on Jefferson and science.

My first duty is to thank Andrew O'Shaughnessy and the entire Monticello staff for their warm and generous hospitality over the past few years, extended both to me and to my wife, Linda. Under his leadership, the center provides an exceptionally supportive environment for scholars. At Monticello, I must also single out for special thanks the staff in the library—Anna Berkes, Endrina Tay, Eric Johnson, and head librarian Jack Robertson—for their aid, along with Jefferson Looney, Ellen Hickman, and Julie Lautenschlager at the Jefferson Retirement Papers Project. I thank others at Monticello—Peter Hatch, Gaye Wilson, Mary Scott-Fleming, Leni Sorenson, Cinder Stanton, Susan Stein, Elizabeth Chew, Sarah Allabach, Sara Bon-Harper, Leah Stearns, and William Beiswanger—who generously shared their expertise and unstintingly helped even when they really did not have time. Leah Stearns helped find images from Monticello for reproduction.

Successive stays at Monticello brought me to meet a whole range of other Jefferson scholars who generously shared both their friendship and their insights into Jefferson, his complex history and his equally complex character. At the risk of offending others by omission, I must particularly thank Andrea Wulf, whose unique viewpoint on the people of the late eighteenth century, the product of her work on horticulture, has been shared in valuable discussions over the past few years. Martin Clagett at the College of William and Mary, author of *The Scientific Jefferson Revealed,* and biographer-in-process of Jefferson's teacher William Small, was a constant source of information difficult to find elsewhere; he has a special gift for ferreting out useful data and an unusual generosity in sharing them.

All historical research depends enormously on the resources of libraries and their staffs, so I am particularly grateful to Martin Levitt, Roy Goodman, Charles Greifenstein, Valerie-Ann Lutz, and Earl Spamer at the American Philosophical Society, my academic home for the past ten years, together with Babak Ashrafi of the Philadelphia Area Center for History of Science. Eileen Matthias, Robert McCracken Peck, Ted Daeschler, and Ned Gilmore at the Academy of Natural Sciences of Philadelphia, Beth Prindle at the Boston Public Library, and Jeremy Dibbell at the Massachusetts Historical Society provided invaluable help, as did the staff of the Small Special Collections Library at the University of Virginia. Jeff Lock, with his knowledge of eighteenth-century instruments, and engineering historian Henry Petroski helped with an enigmatic Jefferson mechanical drawing. I would like to thank the publisher and Jefferson scholar Robert Baron for his encouragement and must mention also the influence of the late Frank Shuffleton through both his fine edition of *Notes on the State of Virginia* and useful conversations.

All historians owe a debt to those who have made the published works of Jefferson, Adams, and Washington available online in searchable form. I laud particularly the University of Virginia's *Rotunda* project and the extensive digital resources of the Library of Congress, the Massachusetts Historical Society, and the Boston Public Library. How quickly we forget how time consuming archival work was only a few years ago.

I owe an additional debt to Andrea Wulf, Martin Clagget, Mary Scott-Fleming, Babak Ashrafi, Andrew O'Shaughnessy, and Linda Price Thomson for critically reading all or part of the manuscript.

The Thomas Jefferson Foundation kindly gave permission to reproduce the sketch of Monticello (Chapter 2) and the photographs of Monticello's front hall, the book stand, and the plow (Chapter 13); the American Philosophical Society, the *Megalonyx* drawing (Chapter 8), the print of the balloon ascent (Chapter 12), and the Indian vocabulary page (Chapter 15); the Academy of Natural Sciences of Philadelphia, the mastodon tooth (Chapter 5) and the engraving of an albino African (Chapter 11); the Division of Political History of the Smithsonian Institution's National Museum of American History, Jefferson's thermometer (Chapter 13); the National Cryptologic Museum, the wheel cipher device (Chapter 14); and the Massachusetts Historical Society, the spherical sundial (Chapter 13). I am grateful to Princeton University Press for permission to quote extensively from their editions of Jefferson's correspondence.

My agent, George Lucas at Inkwell Management, and Yale University Press's wonderful editor Jean Thomson Black (no relation) acted as midwives for the book with seemingly effortless professionalism in sometimes trying circumstances.

Author's Note

The Founding Fathers wrote eloquently and extensively. Well aware of the historical significance of their lives and times, they made copies of their letters; Thomas Jefferson famously used a variety of copying presses and polygraph devices for that purpose. The result is a remarkably complete record of the thoughts of this group of American eighteenth-century intellectuals, lawyers, clerics, and farmers. In the following narrative, wherever possible, I have tried to allow the characters to speak for themselves through their letters and publications. I have also retained their original spelling, punctuation, and emphases, much of which is strange to modern eyes. Jefferson, for example, consistently wrote the possessive "its" as "it's," and John Adams's spelling, especially when he was young, was delightfully erratic. Seeing the texts in this way, while not quite as authentic as reading the handwritten originals, helps capture for us the living person as he sat, perhaps by candlelight, composing works, short or long, to meet the requirements of a particular moment and also knowing that they would probably be read by future generations. It is a privilege to take our turn at finding a new understanding—in this case, to discover Thomas Jefferson's love and use of science.

Introduction

Thomas Jefferson, like Tennyson's brook, goes on forever. There is no stopping him, no coming to an end either of himself as a personality or to the discoveries to be made concerning him. He himself never ceased to grow, and consequently our knowledge of him can never be considered complete.

—E. Millicent Sowerby

JOHN F. KENNEDY, HOSTING AT the White House a dinner for Nobel Prize winners from the Americas in 1962, observed: "I think this is the most extraordinary collection of talent, of human knowledge, that has ever been gathered together at the White House, with the possible exception of when Thomas Jefferson dined alone." It was a precisely crafted tease, discounting his guests and then raising them up, for to be compared with Jefferson, even en masse, was a compliment. And it was also a deft nationalistic touch, pointing out that the intellectual achievements of that one eighteenth-century man from Virginia had never been matched, even two centuries on.

Today, the popular image of Jefferson, the third president of the United States, is rather mixed. He is famous for the Declaration of Independence, his controversial views on slavery, his house Monticello with its complex interpersonal household relations, and such "inventions" as the improved plow moldboard. But we do not always appreciate that in a time of revolution in politics and in the world of the mind Jefferson was always admired (or feared) as a prodigious intellectual—one of the greatest of the age in either America or Europe. And just what that means is often also unclear, especially with respect to his fascination for, and use of, the greatest of modern inventions—science.[1]

Time and history have not dealt easily with Jefferson. He was too mercurial, too complex, and too intellectual even to fit easily in his own time, and it is quite impossible for us to capture him in a few easy phrases.

His place in American history and literature is assured, although he was often out-of-joint with his contemporaries. For many commentators he was a hero, and others dismissed him as a villain. Even by the standards of the day, however, calling him "a mean-spirited, low-lived fellow, the son of a half-breed Indian squaw, sired by a Virginia mulatto father," as a journalist did in 1800, seems intemperate. More politely, John Adams, his crusty on-again, off-again friend, once wrote: "I wish Jefferson no ill; I envy him not. I shudder at the calamities which I feel his candour is preparing for this country, from a mean thirst of popularity, an inordinate ambition, and a want of sincerity." To the Marquis de Lafayette, however, he was "everything that is good, upright, enlightened, and clever, and is respected and beloved by everyone that knows him."

Friend or foe, admirer or detractor, all agree on one thing: Jefferson was a brilliant thinker, at ease in a range of subjects, from the classics (which he read in the original Greek and Latin), international law, history, and the arts to practical agriculture. Guided (famously in one case, at least) usually more by his head than his heart, he could be intransigent over matters that he had thought deeply about. A statesman and family man, leader of men and country farmer, inventor and naturalist, he was dismissive of subjects that he thought a waste of time. A brilliant and intellectually dashing host who set a table with food and wines on a par with those of France, he was happiest when at home on his Virginia mountaintop.

Students of science have not paid a great deal of attention to Thomas Jefferson. Historians of Jefferson have tended to view him as something of a dilettante. He was, to be sure, a wonderful tinkerer, fascinated by inventions of every sort, and his name will (perhaps unfortunately) forever be associated just as strongly with devices like his plow moldboard, various polygraph machines, and the swivel chair as with the history of ideas. He was definitely an architect of distinction. But in the all-too-common view, Jefferson was a typical Virginia gentleman-philosopher, able but not really serious about science, and certainly not a scientist in the sense of actively contributing to the corpus of scientific knowledge.

Jefferson lived, as we do, in times of exciting and often dangerous change, when change was often driven by science. Some of his most significant contributions to science are contained in his book *Notes on the State of Virginia,* written around 1783. *Notes* was very much Jefferson's personal manifesto, an eclectic combination of commentaries on philosophy, law, and what we would call sociology and science. All were his lifelong loves, especially the sciences: "*Nature intended me for the tranquil pursuits of science.*" But when he was asked, after he had retired from the presidency, whether he would write a new version of *Notes,* he declined, saying, "*The work itself is nothing more than the measure of a shadow, never stationary, but lengthening as the sun advances, and to be taken anew hour to hour.*"[2]

It was Jefferson's firm conviction that science was essential to the progress of the nation and that knowledge and understanding were most valuable when they led to practical applications. This view naturally led him to that shifting area of intersection between the pure sciences and technology that we call invention. The reality, however, is that Jefferson's love of nature and his aptitude for mathematics and logic, together with his lawyerly training (and, it has to be said, disposition), naturally drew him also to the deeper, philosophical side of the sciences. This latter aspect of Jefferson's science—less concerned with the concrete and the practical—has been de-emphasized by biographers. It involved the "natural philosophy" that people like Descartes, Galileo, and Copernicus had begun more than a century earlier to discover the fundamental laws that control the operations of the material world. In Jefferson's formative years, the French encyclopedists and the Swedish botanist Linnaeus were organizing knowledge in ways that made it doubly useful to the practical and the theoretical scholar alike, and accessible to many more.

Jefferson used his science, in several interrelated ways. From his vast knowledge of natural history he demonstrated and constantly asserted the superiority of North American wildlife, landscapes, agricultural potential, and natural resources over anything that Europe could offer (and that European scholars might claim). America was not just the *new* world; the rest of the world could also learn from it. Charles Miller has argued persuasively that "insofar as nature symbolized America in its entirety, nature *was* America for Jefferson."[3] America's nature and riches extended over the whole continent and he saw that all the territory from Atlantic to Pacific, being one whole in nature, must be politically and governmentally whole also. Personally, as a deist, he believed that God was to be found through the study of nature and reason (rather than through revelation). More generally, he saw science (which in his day included both natural philosophy and natural history), when partnered with technology and reinforced through education, was essential for a forward-looking young nation, a way to lead it out of the colonial-era wilderness to a future promised land of freedom and prosperity. Americans had to be well educated in the sciences, and they had to be able to translate the disciplines of science into tangible results. (Ironically, although Jefferson believed that the foundation of the country was its farmers, science and education together ended up producing the contradictory result of leading to industrialization.)

From his earliest years as a student at the College of William and Mary in Virginia, Jefferson was a seeker of truths that described and explained the material world and of certainties that would guide his personal and political philosophies. He sought them both in science and in law. Among them were "the laws of nature and of nature's god," as he put it in the Declaration of Independence, and the rights of freedom and equality for American citizens. These truths could have immense power for progress and improvement of people's lives. He was well aware that many people would find a world of change and progress to be a dangerous place. But he also knew that change was inevitable.

Jefferson brought to all this a passion for precision and logic. He always sought out certitude, but fundamental truths were often impossible to ascertain in a time when science (and the intellectual world) was changing rapidly. He sometimes found himself holding to mutually exclusive positions. In that predicament he was not unlike most of us today. His "*shadow*" steadily advancing with the sun was therefore a metaphor for more than his own book; it was science and, indeed, all knowledge. Then and now, as he recognized, "*science is progressive . . . What was useful two centuries ago is now become useless . . . What is now deemed useful will in some of its parts become useless in another century.*" Everything must change. "I am certainly not an advocate for frequent and untried changes in laws and constitutions," he wrote. "But I know also, that laws and institutions must go hand in hand with the progress of the human mind. As that becomes more developed, more enlightened, as new discoveries are made, new truths disclosed, and manners and opinions change with the change of circumstances, institutions must advance also, and keep pace with the times. We might as well require a man to wear still the same coat which fitted him when a boy, as civilized society to remain ever under the regimen of their barbarous ancestors."[4]

In many ways, the *shadow* is also Jefferson himself, admirable and infuriating, understandable and opaque. Two hundred years on, we can ask important questions about his science, and the times in which he lived. Did he make original contributions to the body of contemporary scientific knowledge, and if so, which have had lasting currency? What did he actually invent? How did his science intersect with his religion and with the simple fact of being an American, not a European? How did his science help create his vision of the young nation and its future?

PART ONE

The Young Jefferson

CHAPTER ONE

Lost: One Large Moose

SOMEWHERE IN FRANCE, PERHAPS in Paris, it is possible—just barely possible—that there still exists a set of moose antlers that once belonged to Thomas Jefferson, third president of the United States. They would now be nearly 250 years old, but antlers are made of hard stuff; unless burned in a fire, they should have survived. Where are they? They spanned some four to five feet. The moose they came from had been a very large one, and that was the whole reason they were sent to Paris in the first place.

These antlers had once adorned the head of a moose living in the woods of Vermont. Jefferson, when American minister to France, went to great expense to have them imported to France. When last seen, they were part of the collections of the royal natural history museum, the Cabinet du Roi, in the Jardin des Plantes, having been sent there by Jefferson on October 1, 1787. But they are not there now. Perhaps they were lost, destroyed, or "borrowed" during the upheavals of the French Revolution. Were they ever formally added into the collections? Nobody knows.

The antlers arrived at Le Havre from Portsmouth, New Hampshire, via Southampton, England, in a large crate and were delivered to the Hôtel de Langeac, the rather grand house that Jefferson maintained on the Champs-Élysées. He kept them for only a few days. Although he had spent forty-five English pounds to obtain them, he gave them away almost immediately. Also in the crate were the bones and skin of another American moose, together with "the horns of the Caribou, the elk, the deer, the spiked horned buck, and the Roebuck of America." All were presented to the royal collections; all are now lost. In fact, the moose skin arrived with most of the fur missing and possibly was not worth keeping anyway.

Nothing about Jefferson's moose was quite what it seemed. The antlers did not belong to the same specimen as the skin and the skeleton, which had come from a gigantic moose standing seven feet high at the shoulder. Moose shed their antlers annually, and this one had been shot in the early spring, when the snows were still on the ground and moose were antler-less. So another set of antlers was put into the shipment—still big, but something of a disappointment to Jefferson.

Though sent to the royal collections, the gift was not meant for the king. It was intended for the Comte de Buffon, superintendent of the royal collections and the greatest natural historian of the age. Nor was it simply a scientific gift. The whole assemblage, accompanied by a letter written in full courtly style, was a not-so-subtle scientific slap in the face, a challenge to Buffon's formidable authority.[1] To use a word with (appropriately) French associations, the gift was an act of American chauvinism. Science was being used, not for the first time and certainly not for the last, as a political weapon.

Jefferson had been trying for a year to get various American friends to find, kill, and send to him a notable specimen of a moose. Finally, his old friend General John Sullivan, a Revolutionary War hero and now president (that is, governor) of New Hampshire, took up the challenge. Sullivan arranged for a troop of soldiers to trek west into Vermont to cull the biggest moose they could find. It was slow work; a road had to be hacked through the snowy forest to get the specimen out. It was prepared according to Jefferson's precise, but inexpert, directions. As Sullivan reported, once the moose had been dressed in the field, "the remaining flesh began to be in a state of putrefaction. Every Engine was set at work to preserve the Bones and Cleanse them from the remaining flesh, and to preserve the skins with the hair on, with the hoofs on and Bones of Legs and thighs in the skin without putrefaction, and the Jobb was both Expensive and Difficult, and such as was never before attempted, in this Quarter. But it was at Last Accomplished exactly agreable to Your Directions, except that the bones of the head are not Left in the skin agreably to your Directions, as it was not possible to preserve them in that Connection, but the head of the skin being whole and well dresst it may be Drawn on at pleasure."[2]

Jefferson was no taxidermist, and if he was a little surprised by the sorry state of the specimens when they arrived, he was even more surprised by the size of the bill. By most measures, forty-five English pounds was the equivalent in modern terms to several thousand dollars. At that time, one could buy (and we know Jefferson did) a pair of silver-mounted pistols for one pound, eighteen pence, and a microscope from Dolland's in London for five pounds, five shillings.

Matters were not improved when the shipment was badly delayed; actually, it was left behind on the dock at Portsmouth in March. This contributed to the decay of the specimens, but when the crate arrived at his house, Jefferson eagerly opened it and laid out the contents, showing them off to every visitor, many of whom may have been surprised to see that such a long-anticipated shipment contained a bunch of odoriferous bones. Then, on the cool, rainy last day of September 1787, everything was repacked and sent off to the Jardin de Plantes, where, Jefferson hoped, Buffon would have them "sowed up and stuffed" for display in the museum.

Buffon and his associates, Louis Jean Marie Daubenton and Bernard Germain Étienne de La Ville-sur-Illon, Comte de Lacépède had been forewarned that the gift was coming. Buffon himself, however, was ill and still had not returned from his summer retreat. Lacépède opened the crate, now somewhat battered and foul-smelling, with a mix of curiosity and foreboding. Would Buffon, sometimes a difficult man to work with, be delighted with the contents or furious? He would be well aware of exactly why Jefferson had sent it.

The writings of Georges-Louis Leclerc, Comte de Buffon (1707–1788), were famous around the world. A lot of his work, like his theory about the origin of the earth (which he theorized was as much as seventy-five thousand years old), was controversial, but his encyclopedic masterpiece, *Histoire Naturelle, Générale et Particulière,* begun in 1749 and still unfinished in 1787, was already enshrined as the standard reference work on natural history in Europe and America. It eventually grew to forty-four profusely illustrated quarto volumes describing the natural history of the whole known world. The moose was Jefferson's challenge to the Frenchman's authority, so clearly established in that book. Such a challenge would not have been possible unless Jefferson had been confident of his skill and knowledge as a scientist.

In the ninth volume (1761) of the encyclopedia, Buffon had written a long essay comparing the animals and peoples of the New and Old Worlds —we would call it a biogeography of the Americas. Never modest about his opinions, he categorically stated that America was a dismal and inhospitable place where the indigenous life was sparse and weak compared with that of the Old World. Everything in the New World was smaller, he said, and he also claimed that the people of the New World were inferior in every way to those of the Old. The New World had no large mammals and no civilized people, only an abundance of snakes and noxious insects and a few puny savages. The Americas were too cold and damp to be suitable for any kind of healthy or abundant life.

One titillating sentence of Buffon's had raised eyebrows in America and produced knowing smiles in Europe; Buffon had written that "the savage of the new world . . . is feeble, and has small organs of generation; he has neither hair nor beard, and no ardour whatever for his female."

When copies of the relevant volume of Buffon's *Histoire Naturelle* reached North American shores, most people were too busy with day-to-day survival, not to mention the ongoing difficulties of being under the thumb of King George, to pay much attention to the offending passages. Most people, that is, but not Thomas Jefferson.[3] It was clear to him right away that the Frenchman's calumnies could easily be rebutted scientifically. Jefferson knew that Buffon had never visited America; all his information was based on hearsay. His ideas seemed to reflect a political agenda as much as a zoological one, although his work was not notable for other politically based opinions or, indeed, for many other opinions based on so little factual evidence.

Once the American War of Independence had ended, it was obvious that the French slurs on the American homeland had to be answered before they produced serious political and economic consequences for the fledgling debt-ridden nation. On arriving in Paris in 1784, Jefferson arranged an introduction to Buffon and presented him with a copy of his brand-new book *Notes on the State of Virginia,* in which he answered Buffon's charges point by point. He also presented Buffon with the skin of an American cougar. Buffon was courteous and invited Jefferson to dinner. But he was not to be persuaded by arguments or tables of data. Nor was he impressed by the cougar. Jefferson thought Buffon had been in error by describing the cougar as the same as the animal known as the puma. Buffon (correctly as it happened) stuck to his ground.

In matters like this, Jefferson was somewhat like a bulldog; once he sank his teeth into something, he did not let go. He decided to give Buffon incontrovertible evidence of the majesty and superiority of American wildlife by arranging to have the biggest possible moose sent to Paris. Nothing in Europe was as big and imposing as an American moose. The moose alone would demonstrate the vigor of America. Even better, in his own book, Buffon had made the mistake of confusing the moose ("l'orignal") with the reindeer ("le renne"); a reindeer, Jefferson scornfully noted, was an entirely different species and so much smaller that one could pass under the belly of a moose.

Buffon was not used to being contradicted. And he may have been surprised that Jefferson was so thoroughly versed in science, although the presence in the French capital of Benjamin Franklin, whom Buffon admired, should have signaled to him that North America was not simply the province of rough farmers, part-time soldiers, and small-town lawyers. Buffon would certainly have been surprised that a relatively unknown political figure from Virginia had dared to challenge his views on any scientific subject, let alone that field in which he considered himself expert. But Jefferson took on Europe's foremost living scientist and, he thought, beat him by dint of having data instead of opinions. In either case, win or lose, with *Notes* he established himself as someone who could make original contributions to science. He had become Thomas Jefferson, man of science.

CHAPTER TWO

The Man Who Could Not Live Without Books

Old Master had abundance of books; sometimes would have twenty of 'em down on the floor at once—read fust one, then tother.

—Isaac Jefferson

FOR EIGHTEENTH-CENTURY EUROPEANS, safe in the gilded (if drafty) halls of London and Paris and the celebrated university towns of Edinburgh and Freiberg and secure in their sense of superiority, the idea that Americans could be serious intellectuals was absurd. America was a land of farmers and Indians, forests and swamps, mosquitoes and bears. Even as late as 1820, Sydney Smith, an English littérateur and coiner of witty aphorisms, observed: "In the four quarters of the globe, who reads an American book? Or goes to an American play? or looks at an American picture or statue? What does the world yet owe to American physicians or surgeons? What new substances have their chemists discovered? Or what old ones have they advanced? What new constellations have been discovered by the telescopes of Americans? Who drinks out of American glasses? Or eats from American plates? Or wears American coats or gowns? or sleeps in American blankets? Finally, under which of the old tyrannical governments of Europe is every sixth man a slave, whom his fellow-creatures may buy and sell and torture?"[1]

One exception to this blanket condemnation of America and Americans was Benjamin Franklin, who was admired by the British and adored by the French and to whom the Scottish philosopher David Hume wrote, "America has sent us many good things, Gold, Silver, Sugar, Tobacco, Indigo, etc. But you are the first philosopher, and indeed the first Great Man of Letters, for whom we are beholden to her."[2]

If Franklin was America's first scientific philosopher, then Jefferson was the second. The author of *A Summary View of the Rights of British America,* the Virginia Statute for Religious Freedom, the Declaration of

Independence, and *Notes on the State of Virginia* could justifiably claim to be what the French called a *philosophe* or *savant*—a philosopher and man of wide learning. And that was precisely how Jefferson was seen by the man who later introduced him to Buffon in Paris—another French nobleman, François Jean de Beauvoir, Chevalier de Chastellux. Chastellux had been third in command of the French forces at the Battle of Yorktown, the decisive battle that ended the American War of Independence in 1781. Although he had entered the army at age thirteen, Chastellux eventually became a considerable literary figure in Paris. He stayed on in the United States after the war and visited Jefferson at Monticello in 1782 during a tour of the country. Expecting to meet a politician, he found a philosopher: "A man, not yet forty, tall, and with a mild and pleasing countenance, but whose mind and attainments could serve in lieu of all outward graces; an American, who, without ever having quitted his own country, is Musician, Draftsman, Surveyor, Astronomer, Natural Philosopher, Jurist, and Statesman; a Senator of America, who sat for two years in that famous Congress which brought about the Revolution . . . a Governor of Virginia . . . and finally as Philosopher . . . a gentle and amiable man, charming wife, charming children whose education is his special care, a house to embellish, extensive estates to improve, the arts and sciences to cultivate." During his stay at Monticello, when talk went on late into the night, "sometimes natural philosophy was the subject of our conversation, and at still others, politics or the arts, for no object had escaped Mr. Jefferson; and it seems indeed as though, ever since his youth, he had placed his mind, like his house, on a lofty height, whence he might contemplate the whole universe."[3] Chastellux's surprise that Jefferson had become so accomplished without ever having been to Europe is disingenuously charming. But in some ways he was right. Much of the America that he toured was primitive. The roads were appalling. Since the larger towns and cities were spread out along the Eastern Seaboard, the easiest travel was by water. In the South, there was little industry; the majority of the people were subsistence farmers who had to import basic necessities of life.

In fact, rural America was very much like rural France. What America lacked was a glittering capital like Paris (population around 600,000) or

London (possibly as many as 1,000,000 people) and their attendant slums. Philadelphia, with a population of 25,000 to 40,000 in 1776, is often claimed to be the second-largest city in the English-speaking world, but it was smaller than Edinburgh, Scotland (60,000).

It was certainly possible in those days, as it is now, to live a useful life (in America or Europe) without pondering the meaning of things or knowing what the notable minds of the past and the present had discovered or opined. But Jefferson read widely and thought deeply about the nature and causes of everything—from Newton's new mechanics, Aristotle's logic, and St. Thomas Aquinas's natural law to the foundations of modern political systems, religion, language, the condition of humanity, and the glories of nature. To Chastellux's initial surprise (and perhaps to the later dismay of Buffon), Jefferson belonged to a whole class of Americans among whom education was paramount, learning was expected, the arts flourished and the sciences thrived.[4]

Chastellux's view of Jefferson was confirmed by a young Dutchman who traveled to America as the aide to the Dutch ambassador in 1784. Gijsbert van Hogendorp visited the United States Congress in session in Annapolis, Maryland, in April 1784. By this time, Jefferson's beloved wife, Martha, had died. Van Hogendorp wrote: "Mr. Jefferson, during my attendance at the session of Congress was more busily engaged than anyone. Retired from fashionable society, he concerned himself only with affairs of public interest, his sole diversion being that offered by belles lettres. The poor state of his health, he told me occasionally, was the cause of this retirement; but it seemed rather that his mind, accustomed to the unalloyed pleasure of the society of a lovable wife, was impervious since her loss to the feeble attractions of common society, and that his soul, fed on noble thoughts, was revolted by idle chatter . . . It was he who composed the declaration of independence, whose principles do honor to his judgment and whose style proclaims a man who is familiar with the works of the ancients and the great writers of Italy, France and England. He has the shyness that accompanies true worth, which is at first disturbing and which puts off those who seek to know him. Those who persist in knowing him soon discern the man of letters, the lover of natural history, Law, Statecraft, Philosophy, and the friend of mankind."[5]

Jefferson was born in 1743 on the family farm, Shadwell, in Albemarle County, which was then at the western limit of European Virginia, separated from the endless wilderness beyond by the Blue Ridge Mountains.[6] Hilly and densely wooded, this country produced only slightly less fertile farming country than the coastal plain, where established families like the Carters and Byrds had vast rich plantations.[7] Jefferson's father, Peter Jefferson (1707/1708–1757), owned several thousand acres. Around 1767, Thomas Jefferson, trained first as a lawyer, began to create for himself a special home—Monticello—on a 1,000-acre mountaintop property that his father had acquired in 1735. This was the base from which he eventually controlled three other farms (Tufton, Shadwell, and Lego) of more than 5,000 acres, worked by some two hundred slaves, and another 4,850 acres at Poplar Forest in Bedford County.

In the parlor at Monticello, Jefferson hung portraits of the three men whom he considered his intellectual heroes: Francis Bacon (1561–1626), statesman, philosopher, and scientist; Sir Isaac Newton (1643–1727, the greatest scientist perhaps of all time; and the philosopher John Locke (1632–1704). In a 1594 work, Bacon laid out what it took to be a modern gentleman-scholar. The words could have been written with Thomas Jefferson in mind. "First, the collecting of a most perfect and general library, wherein whosoever the wit of man hath heretofore committed to books of worth . . . may be made contributory to your wisdom. Next, a spacious, wonderful garden, wherein whatsoever plant the sun of divers climate, or the earth out of divers moulds, either wild or by the culture of man brought forth, may be . . . set and cherished: this garden to be built about with rooms to stable in all rare beasts and to cage in all rare birds; with two lakes adjoining, the one of fresh water and the other of salt, for like variety of fishes. And so you may have in small compass a model of the universal nature made private. The third, a goodly, huge cabinet, wherein whatsoever the hand of man by exquisite art or engine has made rare in stuff, form,

This sketch by Jefferson shows an early vision of Monticello, his house in Charlottesville, Virginia. The final appearance of the house was the result of more than twenty years of building and rebuilding. Courtesy of the Thomas Jefferson Foundation, Inc., at Monticello.

or motion; whatsoever singularity, chance, and the shuffle of things hath produced; whatsoever Nature has wrought in things that want life and may be kept; shall be sorted and included. The fourth such a still-house, so furnished with mills, instruments, furnaces, and vessels as may be a palace fit for a philosopher's stone"[8]

Jefferson followed this prescription almost to the letter. Monticello ("little mountain"), built at the summit of an outlier of the Blue Ridge Mountains, was his home for more than fifty years, during which time he constantly designed and redesigned, built and rebuilt, creating an American architectural masterpiece. It was the home he always looked back to, wherever his public duties took him. In addition to the building itself, magnificent for its time and designed by Jefferson himself in the style of Andrea Palladio, he created gardens in which he experimented with wild and cultivated species. He had fish ponds, and he tried to create a park stocked with deer and elk. The huge Entrance Hall of the house became his "cabinet" (museum) full of fossils, paintings, Indian artifacts, and maps; it was dominated by an "engine," a huge double-faced clock, of his own design. He even kept a tame mockingbird. His study was filled with scientific instruments, but unlike Newton, he was always far too rational to be seduced into searching for the "philosopher's stone" that held the secrets of eternal life or could transform lead into gold.

Jefferson was a man of immense gifts and all-too-human flaws. One of the latter was an inability to manage his finances. He was an insatiable buyer of every kind of fine thing, from French wine to English telescopes and microscopes. Above all, he loved—and bought—books. When in Paris in the 1780s, he devoted every afternoon "in examining all the principal bookstores, turning over every book with my own hand, and putting by everything which related to America, and indeed was rare and valuable in every science. Besides this, I had standing orders" in Europe, "particularly Amsterdam, Frankfort, Madrid and London, for such works relating to America as could not be found in Paris."[9]

Today, in the modestly titled Book Room at Monticello, row upon row of books reach up to the ceiling, their leather bindings glowing, the gold titles on their spines promising visions of worlds both familiar and far away. Jefferson's was indeed, a "most perfect and general library," wherein could be found "whosoever the wit of man hath heretofore committed to books of worth." All human learning, it seemed, was captured in this holy of holies, where visitors in Jefferson's day were almost never allowed to venture.

Jefferson's Book Room was a small space, sparsely furnished with a tall reading desk, an octagonal table with file drawers, neatly labeled, and a chair or two, but not much else. Today it is elegant and ordered, but in Jefferson's time—what a mess. Jefferson did not collect books for their fine morocco bindings or rarity. His were working books, some in tattered paper covers, many of them well worn by use. There were so many books that they spilled over into Jefferson's study next door—his private office, already crammed with scientific instruments, paintings and busts of famous men, and, at his writing desk, his famous polygraph device for making duplicate copies of his letters. More piles of books and papers overflowed everywhere in the house.

Jefferson, as he once said, could not live without books. According to a tally he made in 1773, he already owned 1,256; by 1783 the number had grown to 2,640. He bought another 2,000 books when American minister in Paris (1785–1789). Eventually his "great library" totaled 6,487 volumes, representing 4,931 separate titles—a collection that spanned every subject from archaeology to zoology. The number actually sent to Washington, DC, when he sold his library may have been more than 6,700.[10] Collecting books was no idle hobby for him; he read and used all of them. The library was that of a farmer, an architect, a scientist, a musician, an inventor, a lawyer, and a statesman—an intellectual.

Jefferson's library tells us about the man. On reviewing the catalogue, the reader is immediately struck by the variety: it is virtually impossible to find a subject on which Jefferson did not own a book. Few topics existed upon which he could not discourse expertly and at length. He amassed books on church history, geography, sculpture, painting, music, medicine, agriculture, gardening—the list goes on—and owned all kinds of literature, ancient and modern. He read easily in French, Italian and Spanish, as well as Latin and Greek.[11] His library was, in short, a broad sampling of human knowledge and understanding.

Jefferson was not alone in treasuring and collecting books. His sometimes friend, sometimes rival John Adams, left a library of some 3,500 volumes to the town of Braintree, Massachusetts, and that did not include a whole range of novels and other books that went elsewhere. George Washington, a man of action but nonetheless a Virginia gentleman-farmer, had a library of 1,000 or so volumes. Benjamin Franklin and James Madison had around 4,000 books each.

In the second half of the eighteenth century, colleges like William and Mary, Harvard, Yale, King's College (Columbia), the College of New Jersey (Princeton), and countless day schools gained stature, and their students became the backbone of the new republic. Whereas young men had formerly traveled to Europe to study, more and more of them, like Jefferson himself, were now trained at home.[12] In the country that had produced

Benjamin Franklin and Thomas Jefferson, and that was defined by the Declaration of Independence, no one needed to feel inferior in the company of Europeans. The thirteen colonies had many highly literate households. Literacy, in the remotest hamlet, began with reading the Bible. In fact, literacy was an important area of difference between America and Europe, being close to 85 percent in the supposedly backward colonies, 20–25 percent higher than in Europe.[13]

That it was not just the men who read and thought seriously was established forever by the example of Abigail Adams, the brilliant wife of John Adams, a woman of acute political and social perception, a writer of style, and an avid reader. Adams's distant cousin Hannah Adams wrote *A Summary History of New England* in 1799, among other works.

Whether in Boston, Philadelphia, or Williamsburg, men and women of substance read widely, which meant that they created libraries great and small. For example, the *Virginia Gazette* of August 8, 1771, advertised more than two hundred titles for sale at its Williamsburg offices.[14] Among the Virginia gentry and would-be gentry, a fine library was expected as a matter of course, and considerable sums were expended in bringing books from Europe. The patrician William Byrd II of Westover, Virginia, may have had the largest personal library of the time: it had more than 4,000 titles at his death in 1744.[15]

Although libraries were a mark of one's wealth and taste, they were not created for show. For any well-educated family a comprehensive library was a functional asset, especially in rural America, where, if one did not own a particular book, the chances of finding a copy were slim—except by borrowing it from another household. Membership lending libraries like the one Benjamin Franklin founded in Philadelphia in 1731 were few and far between. Whether a person had need of practical information or was following intellectual interests, a working library was as essential as the plow was to the farmer.

Jefferson doubtless took his love of books from his parents. His father, Peter Jefferson, a tobacco planter and surveyor whose wife, Jane, was a member of the influential Randolph family, had a good library that was unfortunately lost in a fire in 1770. Jefferson lamented, "I calculate the *cost* of the books burned to have been about £200 sterling. Would to God it had been the money, *then* had it never cost me a sigh."[16] We do not know just how many books were lost in that fire, but if the library was worth £200, it probably contained four hundred to five hundred works.[17] At least forty of the destroyed books were Jefferson's own.[18] After this loss, Jefferson immediately started a new collection, which eventually became his great library.

In 1815 the loss of the Library of Congress when the British burned Washington, DC, offered Jefferson an opportunity to raise some money—$23,950, to be precise. He sold to the nation the extraordinary library that he had been amassing for forty-five years, and one can only guess what a wrench it was. Fortunately, he did not live to see the majority of his collection, the foundation of the new Library of Congress, lost in a fire on Christmas Eve, 1851. After selling his library, Jefferson immediately started a new collection, together with yet another, smaller one for his estate at Poplar Forest.

The largest proportion of titles in Jefferson's great library—some 800, or fully 16 percent of the total—were what we today recognize as "science" books. They included 18 mathematics books, 150 books on medical subjects, including surgery, 82 books on zoology and botany, 60 books on physics, and 30 books on chemistry.[19] The next largest categories were history and law.[20] Nonetheless, Jefferson had no single category for science. The word "science" did not mean for Jefferson and his contemporaries what it does now. For most eighteenth-century thinkers, it meant both "knowledge" (the original meaning of the Latin word *scientia*) and the system of discovery and analysis that produced knowledge. The general equivalent of today's science was a combination of two distinct subjects: natural philosophy and natural history.

This division of the sciences was set by Francis Bacon in a book that helped launch the Scientific Revolution of the seventeenth and eighteenth centuries: *The Proficience and Advancement of Learning* (1605).[21] If his philosophy of the scientific method could be captured in a single idea, it would be that progress in learning must proceed from disciplined experiment and study, unfettered by preconceptions or the received wisdom of the ages. He emphasized the supreme importance of reasoning, experiment, and the evidence of the senses. Whereas most philosophers had previously thought their goal was to find the ultimate or final causes of phenomena (that is, God's hand or God's purpose), Bacon looked for secondary causes in the immediate properties and laws of material phenomena. He encouraged people to examine the truth discoverable through their own senses instead of relying on the authority of the Bible or theologians: reason, not revelation, was the source of true knowledge. And that shift in the search for truth propelled the European world into the Scientific Revolution and the Enlightenment.[22]

Bacon rejected the traditional deductive (*synthetic*) approach, in which one starts with a premise or prior theory and then finds evidence to support it. Inductive reasoning (*analysis*), Bacon said in an idealized claim, was the reliable path to the truth. It starts with the careful assembly of facts (where possible through experiment), independent of prior conclusions, and builds to a conclusion by assembling the facts, articulating axioms, and then defining explanatory laws. Only when a law has been framed is synthesis brought into play. As one exponent of Newtonian principles explained further, the methodology of natural philosophers is both synthetical and analytical. They deduce "forces of nature" by analysis of specific phenomena and on that basis, using synthesis, show the makeup of other forces of nature.[23]

Bacon divided the world of knowledge into three grand categories according to the ways in which the brain was used to process them. *Memory* was the category for all History, the subjects where the brain recorded and catalogued facts about real things and events. *Poetry,* his second category, included the intuitive and creative arts and literature. *Philosophy,* based on reason and logic, was by far the grandest division in his view, and it comprised both moral and natural philosophy. Natural philosophy encompassed the more fundamental parts of science, those involving the search for the causes and controlling laws of the material world.[24]

Following Bacon, Jefferson arranged his books on Natural Philosophy (physics and optics, mechanics, astronomy, and mathematics) alongside books on Moral Philosophy, Common Law, and Religion.[25] Zoology, Botany, Medicine, and Agriculture were all grouped together in an entirely different section of the catalogue, paired with histories like Ancient History, South American History, and Ecclesiastical History. What might seem to us to be the eccentric organization of Jefferson's library helps us embed Jefferson and his science firmly in the intellectual life of his age, not some diluted version of our own.

CHAPTER THREE

Schooling, Formal and Informal

PHILOSOPHY IN ITS BROADEST SENSE—love of knowledge and inquiry into the basis of knowledge—was in Jefferson's day infused with the living spirit of the Enlightenment. Jefferson and his contemporaries were the direct beneficiaries of recent intellectual achievements ranging from the discovery of basic laws of the universe to the establishment of the law of nations. This kind of philosophy was not only exciting but necessary and immediate, not at all arcane but practical. For men like Jefferson it became the foundation and guide in their private and public lives; they did not acquire knowledge simply for its own sake but as the basis for change in the human condition. Among other things, this kind of philosophy shed new light on the concepts (now so familiar as often to be taken for granted) of life, liberty, and the meaning and possibility of happiness. This kind of philosophy changed the whole world, and science was one of its central elements.

Jefferson began his formal schooling when his father placed him "at the English school at five years of age; and at the Latin at nine . . . My teacher, Mr. Douglas, a clergyman from Scotland, with the rudiments of the Latin and Greek languages, taught me the French . . . On the death of my father, I went to the Reverend Mr. Maury, a correct classical scholar, with whom I continued two years."[1] It was at Maury's school that Jefferson's introduction to formal natural history (and possibly also to natural philosophy in the form of Isaac Newton's celestial mechanics) began.[2]

In 1760, at the age of sixteen, Jefferson began two years of study at the College of William and Mary in Williamsburg, Virginia, a period that coincided with the brief tenure there of William Small. As Jefferson wrote in his *Autobiography,* "It was my great good fortune, and what probably fixed the destinies of my life, that Dr. William Small, of Scotland, was then Professor of Mathematics, a man profound in most of the useful branches of science, with a happy talent of communication, correct and gentlemanly manners, and an enlarged and liberal mind. He, most happily for me, became soon attached to me, and made me his daily companion . . . and from his conversation I got my first views of the expansion of science, and of the system of things in which we are placed."[3]

Small (1735–1775) had studied at Marischal College in Aberdeen, on Scotland's east coast, then a center of teaching and learning in natural philosophy.[4] Isaac Newton's *Principia Mathematica* (1687) had single-handedly reversed two thousand years of thinking about mechanics and the motions of the planets. Small brought Newtonian science to William and Mary and to the young Jefferson. He also brought the special emphasis of the Scottish Enlightenment on the power of reason to discover means of improving nature and society. In Small's lectures (a mode of teaching that he introduced to the college), natural philosophy was both an academic subject and an exciting practical one, yielding a particular worldview of law and order, process and consequence, discoverable through human reason and often revealed through thinking about everyday material objects and phenomena and using (at least to our eyes) humble apparatus like the air pump or Newton's glass prisms. The principle that science must yield practical results and applications stayed with Jefferson all his life.

Franklin had recently published his experiments with electricity and, perhaps even more important, had shown that a "colonial" could be a philosopher and scientist on the world stage. The most dramatic changes in sciences, however, belonged to Newton, whose mechanics showed not only that all motion on earth obeyed the same three, deceptively simple laws but that those laws also explained the mechanics of the heavens.[5] Work that others had done on motion, gravity, and cosmology, from Aristotle and Ptolemy to Descartes, Copernicus, Kepler, and Galileo, had suddenly fallen into place. Newton's apparently simple first law of motion —that any object will naturally move in a straight line at a constant velocity (including a velocity of zero) unless acted upon by an external force—was completely counterintuitive and yet, once grasped, opened up a world of explanations of material phenomena. (Before that, since Aristotle and even earlier, an object was assumed to be normally in a state of rest unless acted upon by a force. The difference is fundamental.)

Most exciting of all, of course, was the famous (if legendary) case of Newton's apple, which, as the story went, led to an explanation of the motions of the heavens. Whether or not Newton actually did observe an apple falling to the earth from a tree and accelerating as it did so, he had the inspiration that the moon must be subject to a similar gravitational force of attraction causing it to "fall" toward the earth. Without a force of attraction, the trajectory of moon across the sky should otherwise be a straight line off into space at a constant velocity. The attracting force of the earth's gravity (and the much lesser attraction of the moon's gravity on the earth) combined to capture the moon in a permanent orbit around the earth. In the same way, the earth and other planets orbit the sun. For many readers, like Jefferson, an added attraction in Newton's works was the elegance of his mathematical expositions, the crowning achievement of which was to show that the force of gravity decreased with the square of the distance between two attracting objects, which mathematically required that planetary orbits be ellipses rather than circles—just as Johannes Kepler had discovered.

Lying behind Newton's science was Bacon's. Bacon's ideas can not only be seen in the formality of Jefferson's library catalogue. They were also a foundation for the life of a man for whom all approaches to natural history and natural philosophy, observation and experiment, fact and hypothesis, were important, and reason and logic even more so. Bacon was Jefferson's introduction to the Age of Reason. While European *rationalist* philosophers like René Descartes, Baruch Spinoza, and Gottfried Leibniz emphasized the view that knowledge had to be acquired through (more or less) pure reasoning, including mathematics, Jefferson was more taken with British *empiricist* philosophers like Thomas Hobbes, John Locke, and George Berkeley, who, following Bacon's approaches, emphasized that knowledge could come only from real experience.

In this empiricist tradition, the path-breaking researches of Isaac Newton on motion, gravity, and optics, took on tremendous significance, far beyond physics (natural philosophy) itself. Newton wrote in *Opticks*, his book on light, "If Natural Philosophy, in all its parts, by pursuing this method, shall at length be perfected, the bounds of moral philosophy will also be enlarged."[6] This was a truly revolutionary thought—that natural philosophy would one day be able to explain human behaviors and interactions, customs, and laws. The open question, the elephant in the room, was, What could natural philosophy reveal about God?

There could hardly have been a more intellectually exciting time to be a student (or, indeed, a teacher). The promises of the Age of Reason, that human intellect—rather than divine revelation—would reveal the mysteries of the universe and perhaps, one day, even the subtleties of human nature seemed to be coming true. Nor could there have been a more perfect match of subject and student in terms of precision of logic, elegance of expression, and breadth of interests. Small did not stay long at William and Mary: the window of opportunity for this all-important interaction between teacher and pupil was extraordinarily short. One can only wonder how Jefferson's intellectual development would have proceeded if the two had missed each other.

No record exists of Small's lectures, but we can partially reconstruct them by looking at other contemporary or near-contemporary teachers and the flood of books on natural philosophy that appeared in the late eighteenth century (Jefferson had sixty or seventy such books in his great library). As a sophomore at Harvard College in the spring of 1754, John Adams (1735–1826), enrolled in Dr. John Winthrop's celebrated course on natural philosophy. Winthrop was the foremost man of science in America after Benjamin Franklin. He became famous for, among other things, pioneering work on earthquakes and sunspots.[7]

Adams, for whom spelling was often a haphazard venture, wrote in his diary: "Mr. Winthrop began a series of Experimental Phylosophy, and in the 1st place he explained to us the meaning, nature, and excellence of natural phylosophy, which is, (he says) the knowledge of those laws by which all the Bodys, in the universe are restrained, it being evident that not only those great masses of matter the heavenly Bodys, but all the minutest combinations of matter in each of them are regulated by the same general laws. For instance it is plain that all the planets observe exactly the same uniform rules in their revolutions round the sun, that every particle of matter observes on the surface of the earth."

The entry continues: "As to the usefulness of natural phylosophy, to be convinced of that, it is necessary only to reflect on the state of all the Civilized nations of Europe, compared to many nations, in affrica, of as quick natural parts as Europeans, who live in a manner very little superiour to the Brutes.—The first Cause, and indeed the alpha and omega of natural phaenomena, is motion, their being an utter impossibility that any effect should be produced in a natural way without motion, and which motion is subject to Certain laws which he explained, and I have forgot. But thus much I remember, that motion, produced by gravity, was universally in right lines, from the body acted upon by gravity, to the Center of gravity, as the Center of the earth, for instance, or the like."[8]

Winthrop's lectures, and undoubtedly Small's also, strongly emphasized the useful and practical. Winthrop opened with lectures on "powers and weights" and centers of gravity and motion and proceeded to simple machines like the lever, pulley, and screw balance and then to complex machines like a crane. This led directly to a treatment of Newton's laws of motion and "the theory of Centirfugal forces . . . aplyed to the Cases of the planets." In just seven lectures, as Adams recorded, Winthrop took his young students from the mundane to the sublime.

Teachers of the time, after an extensive treatment of gravity reflecting Newton's path-breaking work, dealt with the properties of water and air and elements of hydrostatics and pneumatics. Next came electricity. In the hands of a host of experimenters, including Charles-Augustin de Coulomb, who defined the forces of electrostatic attraction and repulsion, Pieter van Musschenbroek, who invented the Leyden jar, which could store an electrical charge, and Franklin, who showed that electricity was a single "fluid" with complementary positive and negative charges (and, of course, that there was electricity in thunderclouds), electricity had been transformed from a parlor trick, demonstrated with a silken dress and a glass rod, to a philosophical subject. Within twenty years, Luigi Galvani would discover the relation between electricity and muscle contraction. Another essential subject taught in a natural philosophy class was optics, the field that Newton had revolutionized with his discovery that white light could be split into colors by a prism and then recombined.[9]

One teaching device of Winthrop's was to have students consider the solar system from the viewpoint of a "Spectator in the Sun, which is of all the most Simple Case." Another was to use an orrery to demonstrate the "Motions & Phaenomina of the Planets & Plantery System." College teaching in natural philosophy was usually accompanied by demonstrations using a variety of mechanical devices. As the subject became more and more popular, there developed a minor industry in creating new equipment for experimentation.

When Jefferson was posted to Paris in 1784, he made a trip across the English Channel to London, where he bought the latest apparatus and equipment for experiment and observation. Jefferson always delighted in all kinds of gadgets, tools, and scientific instruments. He bought them for himself and to give away to his friends. A spendthrift who was constantly in debt, he nevertheless carefully wrote down each purchase in one of his memorandum books. Perhaps, if he had been equally diligent in recording his income and reconciling it with his expenses, he might have remained more solvent.

During the seven-week trip to London, Jefferson made purchases on almost every day except Sundays. What he bought tells us something both about his temperament and interests and about the restrictions of life in America. In Europe, Jefferson found himself in an Aladdin's Cave of sophisticated craftsmanship, especially of scientific items. In London, home of the best instrument makers, Jefferson purchased a telescope, a thermometer, a hygrometer, a camp theodolite, a hydrostatic balance, a scioptic ball,[10] a scale, some double convex lenses, a chest of tools for woodworking and metalworking, two kinds of microscopes, "mathemat. instruments," an "air-pump & apparatus," and a "little electric machine" (a bottle that could be rubbed with a ribbon to create a charge).[11] He was less interested in clothing, but he did buy a new suit, something more appropriate to diplomatic life in Paris than his usual attire. He was no dandy but the English-style frock coat was then all the rage in France. He also had his portrait painted by Mather Brown. And bought a pair of pistols, a sign of the tempestuous times in which he lived.

William Small left the College of William and Mary in 1762, ostensibly to purchase a suite of experimental apparatus for teaching in the college. But he never returned. Instead, he joined the inventor Matthew Bolton in Birmingham, England, and became an important member of the famous Lunar Club, which was so influential in the evolution of Britain's science and industry.[12]

One of the great benefits of learning Greek and Latin (apart from reading the classics in the original) is that the structure of their grammar illuminates the sentence patterns and usages of English and the Romance languages. The benefit of learning philosophy and science (apart from mastery of facts and theories) is an understanding of the processes which create knowledge and test it. Underpinning all this learning is logic. One of Small's lasting gifts to Jefferson was to establish the close relationship between natural philosophy (science) and the rest of philosophy. This was not simply a matter of historical accident or the consequence of adopting Bacon's scheme to organize a library. For Small and Jefferson, the cement binding together the world of learning (including natural philosophy and natural history) was a pair of subjects—*rhetoric* and *logic*—that Small had not expected to teach at all.

The College of William and Mary had only a few faculty members when Small was appointed and was in great disarray when he arrived. The Reverend Jacob Rowe having been expelled from the college for drunkenness and leading the students in a riot, Small, as the only remaining philosopher, was required also to teach moral philosophy, which meant "Rhetorick, Logick, and Ethicks." In fact, according to Jefferson, Small was the first "who ever gave, in that college, regular lectures in Ethics, Rhetoric and Belles Lettres."[13] Small's temporary duties in this regard ended a year later with the appointment of a successor, Richard Graham, in June 1761. The brief period during which Small taught moral philosophy thus coincided with the end of Jefferson's first year and the beginning of his second.

When asked to teach moral philosophy, Small took as his guide the philosopher John Locke's *Two Treatises of Government* and *Essay Concerning Human Understanding.*[14] Locke's "second treatise" concerned "The Original (ie the origins), Extent, and End of Civil Government." In it he proposed a theory of the natural rights of citizens, and that philosophy of individual liberty and equality became the basis of many Americans' views of their rights within, and ultimately against, the British empire. In logic and rhetoric, Small followed the work of Lord Kames and William Duncan. Duncan's ideas— refinements of the classical formalities of Aristotle and Cicero—later became enshrined in the works of many others, from Joseph Priestley to Hugh Blair.[15]

Duncan (1717–1760) approached knowledge from the point of view of explanation; his scheme of knowledge emphasized the ways a scholar must argue a case: both to formulate a result properly and to explain it to others. He emphasized that the appropriate use of reasoning and logic was at the core of any discipline.[16] Without them, one could not properly establish and argue from premises, frame an experiment, test a hypothesis, or even organize and analyze data. The correct use of rhetoric and argument allowed the philosopher to explain, persuade, and promulgate difficult concepts and complicated chains of connectedness.

For Jefferson, rhetoric and logic were a crucial complement to Bacon's teaching on the scientific method itself. He also followed closely Newton's rules for reasoning. The first was, "We are to admit no more causes of natural things than such as are both true and sufficient to explain their appearances," and the second was, "Therefore to the same natural effects we must, as far as possible, assign the same causes." But Jefferson was not just a philosopher-scientist; he also had a lawyer's frame of mind,

Jefferson's training in the law, after he left William and Mary, continued in Williamsburg with the lawyer and patriot George Wythe.[17] It reinforced what he had learned from Small about the value of rigorous logic and precise rhetoric. When occasions came for him to write authoritatively about science, particularly when he analyzed a specific phenomenon, whether the Gulf Stream or the life cycle of the Hessian fly, and when occasions came for him to write as a lawyer and political theorist, the same intellectual discipline and rigor of argumentation shone through.

That Jefferson, all his life, followed the precepts of Bacon and Newton closely is shown in dozens of statements he made during his long career, whenever commenting on a particular subject or a method of approach. In *Notes on the State of Virginia,* irritated by the inaccuracies of statements about America, he pronounced: "Reason and free enquiry are the only effectual agents against error."[18] In correspondence with his friend Charles Thomson about archaeology, he said: "On the antiquities found in the Western country. I wish that the persons who go thither would make very exact descriptions of what they see of that kind, without forming any theories. The moment a person forms a theory, his imagination sees in every object only the tracts which favor that theory."[19] To another correspondent he said, "I am myself an empiric in natural philosophy, suffering my faith to go no further than my facts. I am pleased, however, to see the efforts of hypothetical speculation, because by the collisions of different hypotheses, truth may be elicited, and science advanced in the end."[20] Likewise, when Jefferson was asked for an opinion on meteorites he wrote cautiously. "We certainly are not to deny whatever we cannot account for. A thousand phenomena present themselves daily which we cannot explain, but where facts are suggested, bearing no analogy with the laws of nature as yet known to us, their verity needs proofs proportioned to the difficulty . . . It may be very difficult to explain how the (Meteorites) you possess came into the position in which it was found. But is it easier to explain how (the meteorite) got into the clouds from whence it is supposed to have fallen?"[21]

Jefferson began to have his doubts, however, about the value of studying moral philosophy. He wrote to advise his nephew Peter Carr about his education: "Moral philosophy. I think it lost time to attend lectures in this branch. He who made us would have been a pitiful bungler if he had made the rules of our moral conduct a matter of science. For one man of science, there are thousands who are not. What would have become of them? Man was destined for society. His morality therefore was to be formed to this object. He was endowed with a sense of right and wrong merely relative to this. This sense is as much a part of his nature as the sense of hearing, seeing, feeling; it is the true foundation of morality, and not the truth, &c., as fanciful writers have imagined. The moral sense, or conscience . . . is given to all human beings in a stronger or weaker degree, as force of members is given them in a greater or less degree. It may be strengthened by exercise, as may any particular limb of the body. State a moral case to a ploughman and a professor. The former will decide it as well, and often better than the latter, because he has not been led astray by artificial rules."[22]

The same letter to Carr also reveals Jefferson's feelings about religion. "In the first place divest yourself of all bias in favour of novelty and singularity of opinion. Indulge them in any other subject rather than that of religion. It is too important, and the consequences of error may be too serious. On the other hand shake off all the fears and servile prejudices under which weak minds are servilely crouched. Fix reason firmly in her seat, and call to her tribunal every fact, every opinion. Question with boldness even the existence of a god; because, if there be one, he must more approve the homage of reason, than that of blindfolded fear."

Beyond a library and a fine education, the life of the mind requires interchange with others, and Jefferson is famous for the range and scope of his friendships. Because he did not publish any work in "science" apart from *Notes on the State of Virginia* and a 1799 essay on a fossil ground sloth, it is from his voluminous, catholic correspondence that we can learn the depth of Jefferson's understanding of science and the range of his interests. The subject matter of his letters is as expansive as the catalogue of his library, magnified ten- or a hundredfold.

Within his vast acquaintanceship, Jefferson had a particular circle of friends whose interests in science coincided with his. Together they formed a sort of evolving informal college devoted to both the theoretical aspects of learning and its practical value for America. As Daniel Boorstin wrote in his pivotal study *The Lost World of Thomas Jefferson*, "The Jeffersonian approach to the whole external universe, including society, was pervaded by a kind of integrity and intellectual morality. The desire to know nature was the strongest incentive to ingenuousness, and the most effective restraint against the deception of oneself or of one's neighbours. The Jeffersonian thus attained the 'scientific' frame of mind in the best sense of the word."[23]

The composition of this circle of scientists was constantly changing, and so were Jefferson's ideas and experiences. At first, the group of like-minded men was small, consisting principally of his longtime friend and nephew Peter Carr and the Reverend James Madison (1749–1812, later Bishop Madison and president of the College of William and Mary; cousin of the future fourth president of the United States), plus three older men, Governor Francis Fauquier (1703–1768), a wealthy man of letters and member of the Royal Society of London, his mentor the lawyer George Wythe, and William Small. Jefferson was a keen violinist, and one of the things that brought this group together was music; they played together on a regular basis. Jefferson was also a founding member of the Virginian Society for the Promotion of Usefull Knowledge to Be Held at Williamsburg (1773).[24]

In the years leading up to the American Revolution, while New England patriots like Samuel Adams were building reputations as fiery opponents to British rule, Jefferson was coolly and forcefully using the power of his pen, his command of logic and rhetoric, and his knowledge of the law of nations as a member of the Virginia legislature. The culmination of his thinking at that time appeared in his *Summary View of the Rights of British America* (1774), written as a mandate to the Virginia delegation to the First Continental Congress, which put on the national stage. When he was sent to Philadelphia as delegate to the Continental Congress in 1775, his reputation preceded him. The shy, quiet, almost retiring young lawyer was a natural choice to lead the drafting of one of the defining documents of the new United States—the Declaration of Independence. His place in history was assured from that moment.

In Philadelphia, Jefferson's personal intellectual circle expanded, particularly with respect to science. He now found himself at the intellectual heart of the thirteen colonies. Despite the claims of Harvard College and Boston to be the center of America's elite, scientifically the focal point was the American Philosophical Society, founded in Philadelphia in 1745. The monthly meetings and the published *Transactions* of the society were the principal vehicle for the dissemination of "useful knowledge"—scientific ideas, experiments, and observations by British American intellectuals.

The strength of Philadelphia as an intellectual center depended on the initiative and example of one man: Benjamin Franklin. Franklin (1706–1790) was nearly forty years older than Jefferson and by 1775 had already enjoyed a stellar career with respect to his researches into subjects as diverse as population growth, electricity, and the Gulf Stream. Franklin was also a role model for the application of basic scientific thinking to useful purposes, as with his improvements in the stove and his lightning conductors. Jefferson soon met Franklin and other prominent Philadelphians such as the physician Benjamin Rush, the mathematician and surveyor Andrew Ellicott, the surveyor, astronomer, and inventor David Rittenhouse, and the botanist-explorer John Bartram and his son William. Charles Thomson, secretary to the Continental Congress, an authority on the customs and languages of American Indians, and the first American to translate the Bible into English, became a close friend. With another member of the Continental Congress, Francis Hopkinson (who designed the American flag), Jefferson exchanged information about subjects as disparate as new methods for tuning and quilling harpsichords, ballooning, and patent lamps.

The Continental Congress brought to Philadelphia many of the finest minds from other parts of the country; some of them, like Robert Livingston (1746–1813) of New York had the same sort of intellectual breadth as Jefferson. Livingston shared with Jefferson ideas as disparate as the design of steam engines, the use of steam to pump water to the upper stories of houses in case of fire, and paper making. Prominent among the northerners meeting in Philadelphia was John Adams, who would be the third president of the new republic. Recently arrived in America was also Thomas Paine, an inventor and an architect as well as a pamphleteer and political firebrand.

Jefferson was quickly recognized as a peer among this group, and in 1780 he was elected to the American Philosophical Society, eventually becoming its president.[25] In terms of his growth as an intellectual and a scientist, however, probably the most important single phase of Jefferson's career was the period 1784–1789, which he spent as American minister in Paris, where he circulated among the cream of Europe's intelligentsia. Back in the United States, in the ten years before he became president, Jefferson spent a great deal of time in Philadelphia again, this time as secretary of state. He acquired new scientific colleagues then, including the physician-anatomist Caspar Wistar and the anthropologist and botanist Benjamin Smith Barton. The British preacher and chemist Joseph Priestley and other European exiles, the historian and early geographer the Comte de Volney for one, eventually joined the ever-growing circle of friends, as did the fabled naturalist and explorer Alexander von Humboldt. With all these men and hundreds more, Jefferson maintained a lifelong correspondence.

All the members of Jefferson's circle, at home or abroad, shared one characteristic above all. No matter their (considerable) political and religious differences and regional orientations, no matter the formality or informality of their schooling, all were convinced that for any well-educated man who wanted to achieve greatness for himself and for his country through the application of his intellect, science was a central necessity. Science was not separate from other subjects; it was not a set of ideas or practices that could be picked up or left out at will. For these men, knowledge was indivisible, and all its elements were interdependent, sharing a common philosophical base and a common methodology of reason and logic. That approach to thinking about the world is what separated a true intellectual from the rest.

CHAPTER FOUR

A Measured and Orderly World

ALTHOUGH JEFFERSON LOVED THE philosophical side of science, he remained constantly concerned with its practical application. No fact was too small to catch his attention; what others might find trivial, he turned into useful information. Scattered throughout his notebooks and letters are endless lists of facts and figures and dozens of little gemlike calculations, such as this one testing the efficiency of different kinds of wheelbarrows for hauling dirt. "Julius Shard fills the two-wheeled barrow in 3. minutes and carries it 30 yards in 1½ minutes more. Now this is four loads of the common barrow with one wheel. So that suppose the 4. loads put in the same time viz. 3. minutes, 4. trips will take 4 × 1½ minutes = 6′ which added to 3′ filling is = 9′ to fill and carry the same earth which was filled & carried in the two-wheeled barrow in 4½′. From a trial I made with the same two-wheeled barrow I found that a man would dig & carry to the distance of 50. yds 5. cubical yds of earth in a day of 12. hours length."[1]

It was possibly from his mother, Jane (1720–1776), a member of the prominent Randolph family, that Jefferson acquired or inherited a love of nature and gardens. The Randolphs were keen students of botany.[2] All his life, Jefferson collected natural history information, whether from his farms, from his travels, or from his wide reading. It would not be an exaggeration to say that by 1785 he was, in addition to everything else, America's most knowledgeable naturalist. The gift of Peter Jefferson, his father, was a love of numbers and precision and a passion for recording detailed data and for making sense of it. For Jefferson, knowing and understanding the world required actually measuring it and analyzing the numbers. This applied whether he was recording the weather, which he did wherever he stayed, or using actuarial tables to suggest limits to national indebtedness. The collecting of accurate data, especially where it could be done objectively through instrumentation rather than personal impression, had become an important part of Enlightenment science, with its emphasis on Bacon's methods, and Jefferson reveled in it.[3] Data were not collected simply for their own sake, however; the facts had to be brought together to look for underlying patterns, causes, and laws, especially because, in Jefferson's deist approach to religion, an understanding of the world and nature brought one closer to an understanding of God.[4]

Jefferson learned the skills of the surveyor from his father. Surveying in Jefferson's time involved versions of the same instruments that we use today: a sextant, an accurate chronometer, an accurate compass, a theodolite for establishing levels, and a unit of measure. For Jefferson the measure was a "chain" —a unit of length measuring sixty-six feet (a "rod" was a quarter of a chain). And a surveyor did use a physical chain that long. Armed with these basic devices, two men could walk and measure the dimensions of a plot of land, drive a road through the wilderness, or, in Peter Jefferson's case, lay down the boundaries of a state. On foot, through the wilderness, Peter Jefferson and his neighbor Joshua Fry laid out the Virginia–North Carolina border; in 1751 they were commissioned to produce the first complete map of Virginia.

Thomas Jefferson was fascinated with devices for measuring and observing and devices, like the pedometer and the odometer, that mechanized the tedious measuring process. He wrote eagerly to one of his many correspondents: "I have heard that they make in London an Odometer, which may be made fast between two spokes of any wheel, and will indicate the revolutions of the wheel by means of a pendulum which always keeps it's vertical position while the wheel is turning round and round . . . I will thank you to inform me whether it's indications can be depended on, and how much the instrument costs."[5]

For Jefferson, the quest for precision in all things and his strong mathematical bent allowed him to contribute to a wide range of subjects, including the largest measurements of all. Beyond surveys of the land, he was intrigued by fundamental questions about the shape of the earth and the precise length of a degree of longitude, as in a long memorandum to

James Madison in 1792.[6] He was an avid collector and user of all manner of optical instruments. He owned several telescopes, including those purchased in London, and was a keen astronomer. At the other end of the scale, he had a microscope very early on—there is a record of payment for the repair of one in 1769, and he bought two from the instrument maker William James in London in 1785, a simple "botanical" microscope and a higher-powered compound one. The microscopes were essential to his study in 1790 and 1791 of the life cycle of the Hessian fly. His attention to mathematical precision eventually came into play with his work for standardization of weights and measures, the establishment of a new coinage for the new country, and plans for laying out boundaries in new western lands like "Ohio."

When serving as secretary of state in George Washington's administration, Jefferson was asked to prepare a plan that would standardize coinage, weights, and measures in the new nation. This was indeed essential both for commerce and for science. For coinage, Jefferson proposed, and Congress accepted, a decimal system: the dollar worth one hundred cents. The worth of the dollar, a major topic of debate, was standardized in 1792 as 371¼6th grains (0.085 ounces) of pure silver. For weights and measures the old British system of pounds and ounces, yards, feet, and inches, and acres prevailed. The whole question was perfectly suited to Jefferson's experience and precise turn of mind. Giving some details of what was required will indicate the ease with which Jefferson made mathematics and science useful.

Measure was the most difficult issue, and it was being tackled at the same time in Paris. In Britain the "yard" was legendarily established as the length of the king's arm to the tip of his nose. In the seventeenth century the Royal Society tried to provide a standard in the form of various brass rods machined to a high caliber. But metal expands and contracts with temperature. A more fixed indicator was needed, and the answer (with all its own problems) came from astronomy and surveying.

One fixed quantity in nature was the frequency of rotation of the earth; scientists had already established a "mean solar day." That day could be divided into 24 hours. Given 60 minutes per hour and 60 seconds per minute, the number of seconds in a day could be calculated: 86,400. The unit of time "one second" was fixed by calibrating that unit against the fixed length of a day.

In the seventeenth century, it had been suggested that a fixed unit of length could be established by measuring the beat of a pendulum set to a one-second frequency. (In fact, the period of a pendulum of a given length depends on where on the surface of the non-spherical earth it is measured; length varies because of differences in the earth's gravitational field.) But, as Jefferson told Congress, the pendulum could be "itself a measure of determinate length to which all others may be referred as to a standard."[7]

Newton had already established that, at London, for a one-second beat, a pendulum length of $39\frac{2}{10}$ inches was required. The problem was that setting up and measuring a pendulum was not so easy. If the weight, or bob, hung from a flexible string, the string could bend and shorten the pendulum. The mechanically defined "length" of a pendulum is the distance between the fulcrum and the center of oscillation; for a bob pendulum, measurement is not to the end of the bob, but a point on the suspending string. Instead of a string, the better solution, therefore, was a solid rod. A one-second rod pendulum at London would overall have been $58\frac{11}{16}$ inches in the contemporary English measure.

Jefferson proposed an alternative: that a one-second rod (measured at 38 degrees latitude) should be considered divisible into 587 units, each of which would constitute the now out-of-date dimension of one "line." There were 10 lines to an inch, 12 inches to a foot, and so on.

Jefferson also made proposals for stabilizing liquid measures (pints, gallons), dry measures, and weights—all using the English system of names and relations. Not until 1795 did the French standardize their system of weights on the gram and kilogram, and not until 1797 did they fix the length of the meter at one ten-millionth of a quadrant through the earth at Paris. If those standardizations had been set in place earlier, Jefferson would very possibly have recommended conversion to the full metric system.

It is impossible to tell just when Jefferson's interest in precision and his almost compulsive calculating and note taking began. His preserved Garden Book memoranda began in 1766, his Memorandum Books in 1767, and his Farm Book in 1774. Judging from these, the young lawyer may not have thrown himself into the details of managing the farm and gardens until after his father died in 1767. If any earlier notebooks and diaries existed, they must have perished in 1770 in the fire at Shadwell. It is equally impossible to trace the course of his interests in the more basic aspects of science. He began both a Literary Commonplace Book and a Legal Commonplace Book in his student years or soon after. It seems likely that he would have kept a comparable commonplace book of favorite scientific passages from Bacon or Newton, to say nothing of Aristotle and Pliny. If he did, it is now lost, presumably in the fire.

Jefferson loved making lists. He began his garden diary in 1766 with a short list of the dates at which wildflowers bloomed in the fields around Shadwell, where he lived with his widowed mother. The entries for the following year were far more complete, indicating that he had taken over responsibility for the vegetable gardens. He recorded the first plantings and harvesting of peas (his favorite vegetable), asparagus, celery, onions, lettuce, radish, broccoli, cauliflower, and various flowers, including carnations, roses, sweet william, and lilies. He planted lilacs, roses, and gooseberries and grafted English and black walnuts. He similarly grafted wild and cultivated cherries for his projected new house, Monticello. The garden diary entries for 1767 end with a typical calculation, a hint of many more complex examples to follow: "8 or 10. bundles of fodder are as much as a horse will generally eat thro' the night. 9 bundles × 130 days = 1170 for the winter." When in Washington as third president, Jefferson took the time to chart, weekly, the occurrence of thirty-seven different kinds of vegetables for sale at Washington Market.

Farmers then, as now, paid close attention to the weather, but they did not measure and keep accurate data, usually only recording qualitative assessments such a "warm," "very hot," "dry," or "deep snow." George Washington, for example, kept a detailed diary of such observations; he did not record actual temperatures until 1785. In fact, thermometers were rare in colonial and revolutionary America. Daniel Gabriel Fahrenheit (1686–1736), a German physicist, had not invented the alcohol thermometer until

1709; it was followed by the mercury thermometer in 1714. In the 1760s and 1770s thermometers were obtainable only from Europe; a number of Italian glassblowers had taken the skills with them to London and manufactured them there. The Virginia plantation owner Landon Carter owned a thermometer as early as 1766 but did not use it to keep records. It may well have been broken.[8] The Reverend James Madison at William and Mary College also had a thermometer early on and kept detailed records.[9]

Not always dependent on instruments, Jefferson noted carefully the first dates of various natural phenomena such as the flowering of a particular plant or the arrival of a migratory bird. In 1790, when he was secretary of state and based in New York, he wrote to his daughter Maria at Monticello: "We had not peas nor strawberries here until the 8th day of this month. On the same day I heard the first whip-por-will whistle. Swallows and martins appeared here on the 21st of April. When did they appear with you? And when had you peas, strawberries, and whip-por-wills in Virginia. Take note hereafter whether the whip-por-wills always come with the strawberries and the peas."[10] The reply came back that "we had peas the 10th of May and strawberries the 17th of the same month. As for the martins and whip-por-wills, I was so taken up with my chickens that I never attended them."[11]

The following year Jefferson wrote to Maria: "On the 27th. [of February 1791] I saw blackbirds and Robinredbreasts, and on the 7th of this month I heard frogs for the first time this year. Have you noted the first appearance of these things at Monticello? I hope you have, and will continue to note every appearance animal and vegetable which indicates the approach of Spring, and will communicate them to me. By these means we shall be able to compare the climates of Philadelphia and Monticello. Tell me when you have the pease &c up . . . when you have the first chickens hatched, when every kind of tree blossoms."[12] His conclusion was that the seasons in New York were some four weeks delayed compared to Virginia.

Once he had acquired his own instruments, keeping weather records became almost an obsession for Jefferson, and the data he massed over a lifetime allowed him to make original contributions to American meteorology. His encyclopedic practical knowledge of America, including his many insights into natural history, geography, and climatology, was poured into the book that instantly became a classic and remains so today. That book was the modestly titled *Notes on the State of Virginia.* It began almost by accident.

When Chastellux visited Monticello in 1782, it was a happy time for Jefferson; his circumstances seem to have been much improved over the previous year, when the British had routed him from his home, destroyed his property, drunk his precious wines, and generally embarrassed him in the eyes of his fellow Virginians. The only heroic event associated with the raid by the British under Colonel Banastre Tarleton had been the furious ride of Jack Jouett, a militiaman-farmer from Charlottesville who galloped up the hill to warn Jefferson, then governor of Virginia, who delayed until it was almost too late before escaping southwest to Poplar Forest, his farm in Bedford County.

Now, with the British having been defeated at Yorktown and the war at an end, Jefferson had given up the governorship of Virginia and settled down at Monticello, where his beloved wife, Martha, was expecting their sixth child (of whom, sadly, three had already died).[13] Yet within months Jefferson's rural idyll would be devastated. The new baby died, and shortly afterward, so did Martha; she was not quite thirty-four years old. One of the ways that the bereft Jefferson coped with the twin tragedies was to return to a project that he had begun a year earlier but with which he was dissatisfied. It was perfectly suited to him, combining as it did precision and natural history with rhetoric.

Late in 1780 he had received a request from the French government for up-to-date information about North America. The French had in mind particularly the laws and political customs of the thirteen states that were closing in on independence from the British empire. The secretary to the French delegation in Philadelphia, François Barbé-Marbois, had been directed by his government to write to authorities in each state requesting answers to a series of twenty-two questions. The French wanted to know about ports, militias, natural resources, laws, institutions, and peoples. They even inquired about the ways Americans were dealing with "the Estates and Possessions of the Rebels commonly called Tories."[14] The inquiry for data about Virginia had been sent to Joseph Jones, secretary to the Virginia delegation to Congress, who realized at once that the best person to answer would be Jefferson.

The choice was felicitous. When a fall from a horse required Jefferson to rest, he started writing a response for the French. A compulsive note taker, it was almost as if he had all his life been collecting just the sort of information that Barbé-Marbois needed. "I had always made it a practice, whenever an opportunity occurred of obtaining any information of our country, which might be of use to me in any station, public, or private, to committing it to writing."[15]

Jefferson used the opportunity to write a short report at first. Even so, at an estimated thirty-five to forty manuscript pages, it was five to ten times longer than any of the other reports that Barbé-Marbois received.[16] The respondents for other states did not have the wealth of information that Jefferson had. To be fair, they also did not have the leisure to write long responses to foreigners' questions. Of the other known responses to Barbé-Marbois, Thomas Bee of South Carolina and the Reverend John Witherspoon of New Jersey ignored natural history. General John Sullivan of New Hampshire wrote proudly: "Perhaps few countries have Such a variety of animals for beside all kinds of European Animals moose Elks Deer Bears wolves Catermounts Foxes hares beaver rabbits otters minks Racoons Squirrels & other wild Quadrupeds are found in greater abundance here than in any other country—wild fowl are also found here in very great abundance our Seas Rivers & Lakes abound in Fish of almost Every sort."[17] Only Jefferson answered in detail.

Soon after he had sent his manuscript off to Barbé-Marbois, Jefferson began to enlarge it to book length. Among other things he expanded his view of the glories of America—its natural resources, sublime landscapes, and strong indigenous peoples—showcasing the customs and laws of the new republic. He ended up summarizing the natural history not just of Virginia but of what was then known of North America.

The result was *Notes on the State of Virginia.* In part it is a scientific treatise, and it is often described as the foremost book of natural science in America of the eighteenth century. It was all that, and it was far more.

Beyond the formal descriptions of the laws and conventions, geography, and commerce of Virginia that he had sent to Barbé-Marbois and a comprehensive compendium of natural history information (including geology, animals, plants, fossils, anthropology, and archaeology), it was a personal manifesto. In addition to presenting facts and extolling the virtues of his home state, *Notes* is a declaration of Jefferson's personal philosophy.

Early copies of his draft text were circulated to friends in manuscript form, and the more he wrote, the more worried Jefferson became that the frank expression of his views—especially those on slavery, religion, and the Virginia constitution—would be so unpopular that he would have to keep the project private. Eventually, however, the work had grown to such a size that it could not practically be reproduced longhand; printing copies was necessary. A first, informal edition (lacking even a title page) of two hundred was published in France in 1785; a slightly revised version was published in London in 1787.[18] *Notes on the State of Virginia,* where Jefferson showed the writer's talent that had already been apparent from his *Summary View of the Rights of British America* and the Declaration of Independence, has remained in print ever since.

The book stands in a long Virginian tradition of promotional literature. Possibly the first was Thomas Harriot's *A Briefe and True Report of the New Found Land of Virginia, of Its Commodities, and of the Nature and Manner of the Naturall Inhabitants* (1588). After that came John Clayton's *An Account of Several Observables in Virginia* (1684), Robert Beverley's *The History and Present State of Virginia* (1705), and William Byrd's *The Natural History of Virginia; or, The Newly Discovered Eden* (1737). All of these, together with several in languages other than English, were intended to attract settlers to the New World.

Although much of Jefferson's style of scientific thinking centered on his reading of English natural philosophy from his days at William and Mary College, much came also from two French sources. One was Denis Diderot's *Encyclopédie* and its successor, the *Encyclopédie Méthodique* by Diderot and Jean le Ronde d'Alembert, which summarized for the first time a vast range of human knowledge and introduced Jefferson to such subjects as ancient civilizations and modern technology. He recommended these works to all his friends and, when in Paris, arranged subscriptions for them. The second work, just as large in scope, was Buffon's monumental *Histoire Naturelle, Générale et Particulière,* published between 1749 and (posthumously) 1809.

An important part of Jefferson's approach to science, especially in *Notes,* was framed in reaction against Buffon. Buffon's work was a mass of both facts and theories, mostly sober, a few fanciful. It was all elegantly written, but because it was produced over so many years and with several collaborators, and because Buffon continued to feed into it new interpretations, it was sometimes self-contradictory. Many readers found it glib. Jefferson, who was no slouch when it came to eloquence, especially the elegant dismissal, commented: "I am induced to suspect, there has been more eloquence than sound reasoning displayed in support of this theory; that it is one of those cases where the judgment has been seduced by a glowing pen: and whilst I render every tribute of honor and esteem to the celebrated Zoologist, who has added, and is still adding, so many precious things to the treasures of science, I must doubt whether in this instance he has not cherished error also, by lending her for a moment his vivid imagination and bewitching language."[19]

The theory in question was, for Americans, the most objectionable element of Buffon's *Histoire Naturelle*—namely his derogation, expressed in that alarmingly bewitching language, of the wildlife and the indigenous peoples of the Americas. Buffon's position was that "America" was a uniformly cold and damp place, inhospitable to life, where humans and beasts were sparse and feeble. *Notes* gave Jefferson the opportunity to reply at length; he took on Buffon on his own terms and, as it were, on Jefferson's home ground. It required supreme confidence to do so. *Histoire Naturelle* was the most complete and most influential work on world natural history available to any reader in any language. Whether it was in Buffon's catalogues of the world's mammals and birds, his views on climate and civilization, or his theories of the origin of the earth itself—in no instance could Buffon could be ignored.

PART TWO

Natural Science

CHAPTER FIVE

Science and the Mastodon

ON JUNE 8, 1784, EZRA STILES, president of Yale College, noted in his diary that Thomas Jefferson "is a most ingenuous Naturalist and Philosopher, a truly scientific and learned Man, and every way excellent. He visited the College Library and our Apparatus. Govr. Jefferson has seen many of the great Bones dug up on the Ohio. He has a thighbone *Three Feet long* and a Tooth weighing *sixteen Pounds.*"[1]

Along with measuring things and keeping tables of data, Jefferson loved fossils, especially those of the American mastodon, with its giant teeth, bones, and tusks. The mastodon was a giant American relative of elephants, but just what kind of elephant was unclear; it was often called the American incognitum—the unknown. When Jefferson wrote about American wildlife in *Notes on the State of Virginia,* the mastodon was the first scientific subject he tackled.

Jefferson was a born teacher and, like some teachers, could be an infuriating know-it-all. Just as he enjoyed advising young people on what books to buy and what subjects to study, he enjoyed explaining. As a master of lawyerly argument, he could use his rhetorical skills to demolish a bad idea and promulgate a good one (and sometimes vice versa). His passages on the mastodon in *Notes* are among his most decisive writings on any scientific subject.

Fossil remains of the mastodon, a creature that (as we now know) once roamed across the American interior, had been found in several places in North America since at least 1705, when bones, tusks, and teeth were dug up near the Hudson River in what is now New York State. The Reverend Cotton Mather sent a secondhand report on them to London's

Royal Society in 1714.[2] In the decade before Jefferson was born, mastodon remains were discovered in large numbers at a site just south of the Ohio River, across from present-day Cincinnati. The place was called, appropriately enough, Big Bone Lick. It consisted of a series of marshy springs where, since prehistoric times, animals had come for the salt. Many had become mired in the mud and were eventually entombed and preserved. Giant tusks measuring up to nine or ten feet long were found in the marsh, as were huge teeth, some of which weighed ten or eleven pounds (although most were in the region of three or four pounds). The thigh bones of the animal were three or four feet long, long and strong enough to support three seated men.[3]

The history of the first European discoveries at Big Bone Lick has been told many times before.[4] The first specimens from Big Bone Lick were found by a party of French soldiers in 1739 and were taken back to Paris, where they entered the royal collections and were studied by several notable French scholars. They were described as being the remains of a kind of elephant, but all commentators noted that the teeth were different from those of either the living African and Asian elephants or the Siberian mammoth, whose remains were then being found in the frozen muds of the far north of Europe and Asia. Nothing quite like these teeth from the Ohio Valley was known from the Old World.

The Shawnee Indians sometimes set up camps at Big Bone Lick and boiled down the salt for trade. Their name for the mastodon was the "great buffalo," and Indian wisdom held that the monstrous animals still lived somewhere out in the vast lands beyond the eastern mountains. Perhaps they were right; travelers to the West, who came back with all sorts of improbable stories, like the existence of a whole mountain made of salt, also reported the presence of monsters. They heard strange howlings in the night, and they, too, found giant bones on the ground.

After the French and Indian War of 1754–1763, British American traders and land speculators pushed west along the Ohio. George Croghan collected specimens at Big Bone Lick in 1765. He was ambushed on the way home and lost everything, and then escaped captivity with a tomahawk

This mastodon tooth was in Jefferson's personal collection. He gave it and the rest of his fossil collection to the American Philosophical Society. It was subsequently transferred to the Academy of Natural Sciences of Philadelphia. Courtesy of the Academy of Natural Sciences. © T. Daeschler/VIREO.

wound to the head. Nothing daunted, he made a second trip in 1766 and brought back specimens that were sent to London for Lord Shelborne (minister for the Southern Department) and Benjamin Franklin (who was in London as Pennsylvania's representative). These specimens were studied by such prominent English scholars as William Hunter and Peter Collinson. By the 1780s tusks and more of the strange teeth had found their way into European and American collections.

Whatever the incognitum was, it had been (and possibly still was) a very large animal, perhaps a ferocious one, both to be feared and to be held up to European skeptics as a symbol of the might of American nature. The teeth, with their shiny, dark, coffee-colored enamel, had a strange beauty of their own. The identity of the animals to which the remains belonged remained unclear, however; the world's best brains—even the great Comte de Buffon—had not been able to resolve the nature of this mysterious animal.

The incognitum was just the sort of subject to attract all of Jefferson's scientific and patriotic instincts. It allowed him to answer hearsay with fact, and hypothesis with logic. His treatment of the mastodon wonderfully demonstrates the way he approached science and the logic and rhetoric of writing about science.

One conclusion seemed obvious: those tusks surely were from an elephant of some kind. Both the African and the Asiatic elephants were quite well known by then, and for some years, travelers to the north of the Old World had been bringing back the tusks and bones of Siberian elephants, the "mammoths" that had been buried in frozen mud, sometimes with skin and even meat intact. The ivory of the tusks of the living elephant, the mammoth, and the incognitum was identical. The American remains were from a far larger animal, however—perhaps 70 to 80 percent bigger than a living elephant.

Buffon originally thought that the African and Indian elephants, plus the mastodon and the Siberian mammoth, made up a single species. But, while the American mastodon's ivory was indeed ivory, the teeth were all wrong. All the known "elephants," both the living forms and the fossil mammoths of Siberia, had teeth with multiple low transverse ridges for grinding plant food. The teeth from Big Bone Lick and other sites in North America had five or six prominent blunt conical cusps in two rows. They were also far larger than any known elephant teeth. Those cusps looked for all the world like the dugs of a pregnant dog or sow, and that was the source of the animal's eventual name—mastodon ("breast teeth," as in *mastectomy,* "breast removal")—given by Buffon's successor Georges Cuvier, in Paris, partly on the basis of specimens that Jefferson sent there.[5]

The most obvious conclusion was that the teeth and tusks (principally from Big Bone Lick) belonged to the same species, which was therefore a new, very different kind of elephant, with teeth specialized for eating a different food. That was the conclusion of Britain's leading anatomist, William Hunter.[6] French scientists, however, starting with Buffon's assistant,

Louis-Jean-Marie d'Aubenton, favored the view that the teeth were not from an elephant at all but from a gigantic hippopotamus. From London, Peter Collinson offered a compromise view, suggesting to Buffon that the animal had the tusks of an elephant and the grinders of the hippopotamus.

The ever-astute Benjamin Franklin recognized both the problem and the possibilities when he first saw the teeth of the incognitum in London. He summed up the issues in a letter he wrote to his friend the Abbé Jean-Baptiste Chappe d'Auteroche in Paris. Chappe was an astronomer and, having traveled in Siberia, knew about the European mammoth. Franklin wrote: "Some of Our naturalists here . . . contend that these are not the Grinders of Elephants but of some carnivorous Animal unknown, because such Knobs or Prominences on the Face of the Tooth are not to be found on those of Elephants, and only, as they say, on those of carnivorous Animals. But it appears to me that Animals capable of carrying such large heavy Tusks, must themselves be large Creatures, too bulky to have the Activity necessary for pursuing and taking Prey, and therefore I am inclin'd to think those Knobs only a small Variety. Animals of the same kind and Name often differing more materially, and that those Knobs might be useful to grind the small Branches of Trees, as to chaw Flesh. However, I should be glad to have your opinion, and to know from you whether any of the kind have been found in Siberia."[7]

Elephants are plant eaters. So are hippopotami, although, with their large canine fangs, they were then thought to be carnivorous or omnivorous. The possibility that the incognitum might have used its teeth for chewing flesh rather than macerating plant matter very much added to the mystery surrounding the fossils. There was something titillating about the possibility that a flesh-eating predator larger than an elephant had once roamed the American North. Even more dramatically exciting was the possibility that the animal still lived there.

Although the hippopotamus hypothesis may seem fanciful, d'Aubenton and Buffon were led astray by the some small square molar teeth in the Big Bone Lick deposit that had only four main cusps instead of five or six and did look rather more like hippopotamus teeth than elephant teeth. Buffon described these cusps as resembling the "spade painted on cards."

They have, he said, "every mark of the *dentes molares* of the hippopotamus." (In fact, they were juvenile mastodon teeth.) So Buffon came to the conclusion that the Big Bone Lick site contained remains of three distinct kinds of animals: an elephant (the owner of the tusks) that was the same as the African and Indian elephants, a hippopotamus (bearing the smaller teeth), and the animal to which belonged the "enormous teeth, the grinding side of which is composed of large blunt points."[8]

In Buffon's view, therefore, "independent of the elephant and hippopotamus whose relics are equally found in the two continents, another animal, common to both, has formerly existed, the size of which has greatly exceeded that of the largest elephant . . . This ancient species . . . ought to be regarded as the largest of all terrestrial animals." He did not know what this monster was, stating, "It seems . . . to be certain, that these large teeth have never belonged either to the elephant or to the hippopotamus." But, he hinted, "they have really belonged to a land animal, whose species made a nearer approach to that of the hippopotamus than to any other."[9]

There was a second question, regardless of which species, or how many species, the remains represented: What was any kind of elephant or hippopotamus doing living on the banks of the Ohio, where none live today? Neither of the modern elephant species could possibly live in such northern, seasonally frigid regions. The same difficulty applied to mammoths in Siberia and to the rhinoceros remains found with them. The simplistic explanation was that these were all the remains of tropical creatures that had been washed north in the great biblical Flood of Noah. That meant the unlikely transport of carcasses of drowned animals over thousands of miles. Equally improbable was the idea that they had been taken north, alive, by humans, like the elephants marched by Hannibal across the Alps.

Jefferson had little use for the idea that the Big Bone Lick teeth came from a hippopotamus. "It is remarkable that the tusks and skeletons have been ascribed by the naturalists of Europe to the elephant, while the grinders have been given to the hippopotamus, or river-horse. Yet it is acknowledged, that the tusks and skeletons are much larger than those of the elephant, and the grinders many times greater than those of the hippopotamus, and essentially different in form. Wherever these grinders are found, there also we find the tusks and skeleton; but no skeleton of the hippopotamus nor grinders of the elephant." Buffon evidently had failed to see the flaw in his three-species argument. He had put himself with the position of saying that some elephants had lived at Big Bone Lick and left behind the remains of their tusks and limbs but apparently had had no teeth. This point was not lost on Jefferson, who used pithy sarcasm to make what can be seen as a pretty good joke for a philosophical work: "It will not be said that the hippopotamus and elephant came always to the same spot, the former to deposit his grinders, and the latter his tusks and skeleton." Instead, "We must agree then that these remains belong to each other, that they are of one and the same animal, that this was not a hippopotamus, because the hippopotamus had no tusks nor such a frame, and because the grinders differ in their size as well as in the number and form of their points."[10]

Adopting the style of the courtroom to that of philosophical discussion, Jefferson went on the tackle the question of whether the "mammoth" (the name he always used for the "American incognitum") was "an elephant," by which he meant the living African or Asian elephant. "That it was not an elephant, I think ascertained by proofs equally decisive . . . The skeleton of the mammoth (for so the incognitum has been called) bespeaks an animal of five or six times the cubic volume of the elephant, as Mons. de Buffon has admitted . . . The grinders are five times as large, are square, and the grinding surface studded with four or five rows of blunt points: whereas those of the elephant are broad and thin, and their grinding surface flat . . . I have never heard an instance, and suppose there has been none, of the grinder of an elephant being found in America . . . From the known temperature and constitution of the elephant he could never have existed in those regions where the remains of the mammoth have been found." On the crucial question of climate and environment, Jefferson said, "The elephant is a native only of the torrid zone and its vicinities: if, with the assistance of warm apartments and warm clothing, he has been preserved in life in the temperate climates of Europe, it has only been for a small portion of what would have been his natural period, and no instance of his multiplication in them has ever been known. But no bones of the mammoth, as I have before observed, have been ever found further south than the salines of the Holston, and they have been found as far north as the Arctic circle."

Jefferson then explored the various contemporary theories and their logical consequences. First, those who thought that the mammoth (i.e., mastodon) was the same as an elephant, must also believe that "the elephant known to us can exist and multiply in the frozen zone." As that seemed impossible, then the earth must once have been warmer in the north. Perhaps "an internal fire may once have warmed those regions, and since abandoned them, of which, however, the globe exhibits no unequivocal indications." The third possibility, suggested by European scholars, was that the "obliquity of the ecliptic, when these elephants lived, was so great as to include within the tropics all those regions in which the bones are found; the tropics being, as is before observed, the natural limits of habitation for the elephant." But in that case, assuming that the obliquity has changed at the extreme rate of one minute a century, "to transfer the northern tropic to the Arctic circle, would carry the existence of these supposed elephants 250,000 years back . . . Besides, though these regions would then be supposed within the tropics, yet their winters would have been too severe for the sensibility of the elephant. They would have had too but one day and one night in the year, a circumstance to which we have no reason to suppose the nature of the elephant fitted."

The answer, for Jefferson, was obvious. "For my own part, I find it easier to believe that an animal may have existed, resembling the elephant in his tusks, and general anatomy, while his nature was in other respects extremely different." And he put the question into a geographical framework. "From the 30th degree of South latitude to the 30th of North, are nearly the limits which nature has fixed for the existence and multiplication of the elephant known to us. Proceeding thence northwardly to 36½ degrees, we enter those assigned to the mammoth . . . Thus nature seems to have drawn a belt of separation between these two tremendous animals, whose breadth indeed is not precisely known, though at present we may suppose it about 6½ degrees of latitude; to have assigned to the elephant the regions South of these confines, and those North to the mammoth."

Then, being Jefferson, he ended with a homily to reinforce his view that the fauna of American mammals was not inferior to that of Europe. A direct contradiction of Buffon, it was couched in diplomatic language but nonetheless carried a sting: "But to whatever animal we ascribe these remains, it is certain such a one has existed in America, and that it has been the largest of all terrestrial beings. It should have sufficed to have rescued the earth it inhabited, and the atmosphere it breathed, from the imputation of impotence in the conception and nourishment of animal life on a large scale: to have stifled, in its birth, the opinion of a writer, the most learned too of all others in the *science* of animal history . . . that in the new world, nature is less active, less energetic on one side of the globe than she is on the other. As if both sides were not warmed by the same genial sun" (emphasis added).

These samples of Jefferson's writing and arguments illustrate his deep knowledge of natural history. They show that, for him, "understanding" depended on careful organization of facts, rigorous logic, and an analysis of the correlated consequences of particular facts and interpretations. There was an expectation that nature was orderly and operated according to fixed laws. Above all, precision of expression was required. In a literary sense, the passages quoted are wonderful also because of the way in which Jefferson subtly invites the readers to distance themselves from a renowned authority on natural history, the Comte de Buffon—a denigrator of America. Science, here, was not just something one did for the sake of pure knowledge: it was also a political tool.

In 1784, when Jefferson met with President Ezra Stiles of Yale, he had only two mastodon specimens. The thigh bone was collected for him at Big Bone Lick by George Rogers Clark, who promised, but did not deliver, more remains. The tooth was a gift received two years earlier. Jefferson persuaded the American Philosophical Society to set up a committee to encourage the discovery and study of the "Antiquities of America." The committee, which included Caspar Wistar; the artist, museum director, and entrepreneur Charles Willson Peale; and Jefferson himself, issued a Circular Letter in 1797 setting out priorities. The first objective was "to procure one or more entire skeletons of the Mammoth, so-called, and such other unknown animals as have been, or may hereafter be discovered in America."[11]

Around the turn of the eighteenth century, specimens started to turn up at Claverack, in Columbia County, New York. Jefferson eagerly discussed with Wistar the chances of getting some of the material and wrote to Robert Livingston, a signer of the Declaration of Independence with whom he exchanged correspondence on many scientific issues. Livingston replied that the local townspeople would be unlikely to release any specimens.[12]

Meanwhile, in 1799, news came to Philadelphia that remains had been found on the farm of John Masten, near Newburgh, New York. The lucky recipient of this intelligence was Charles Willson Peale, who was curator of the collections of the American Philosophical Society. He had exhibited mastodon teeth at his new museum, then housed at the American Philosophical Society. With a showman's instinct and savvy, Peale seized his chance and in 1801 began excavations, the most promising of which were on nearby farms.

Jefferson, now president of the United States, offered the loan of a naval pump to keep water out of the marl pit in which the bones were found. But Peale did not want to be beholden to the government. The American Philosophical Society advanced him five hundred dollars, and he succeeded brilliantly in collecting parts of three different mastodons. From these he was able to assemble two more or less complete specimens (with just a few replacement bones carved out of wood). Exhibited in Philadelphia, Peale's mastodon became a sensation.[13]

In 1803, William Clark—the Clark of the Lewis and Clark Expedition and the brother of George Rogers Clark—stopped to collect fossils at Big Bone Lick en route to the West, but found that the site had been denuded by fossil collectors. In 1807 he returned and made an excavation for Jefferson, assembling a large number of bones, sensibly concentrating on collecting the smaller bones that were less commonly collected. He divided the collection into two lots for separate shipment down the Ohio River to the Mississippi and, via New Orleans, to Washington, DC. This was also wise; one of the shipments was lost. When the other, consisting of three hundred bones of half a dozen species of mammals, arrived in Washington, Jefferson laid them out in the vacant East Room of the new White House. Wistar came down from Philadelphia to identify them and to divide them into three lots: one for the American Philosophical Society, one for Jefferson personally, and one that was donated to the Museum.

Jefferson put the more spectacular pieces of his now considerable collection of mastodon remains on display in the White House and at Monticello. Most of those are now lost. The American Philosophical Society's collections eventually went to Philadelphia's Academy of Natural Sciences, where they have been kept safe ever since, and the Paris Museum still has its portion of the collection.

After 1807, discoveries of new mastodon bones continued apace; eventually bones were found in thirteen states. The true mammoth was also found at many sites, including Big Bone Lick. Mastodon specimens have been found with the stomach contents intact; indeed, the animal was a vegetarian, feeding on leaves and brush. No remains of a fossil hippopotamus have ever turned up in the New World, but specimens of the mastodon have been found in South America.

Although Jefferson had written more cogently on the mastodon than had any of the scientists of Europe, he was never sure that the mastodon was a different species from the mammoth, even after the difference between the two species had been accepted by all other authorities. And to the end of his life he persisted in believing that the mastodon was not extinct but would be found living, as Shawnee legend said, somewhere in the great western wilderness.[14]

CHAPTER SIX

The Natural History of Virginia and America

THE AMERICA THAT JEFFERSON described in *Notes on the State of Virginia* was a place of real and promised riches. In the words of his friend Charles Thomson, "This Country opens to the philosophic view an extensive, rich and unexplored field. It abounds in roots, plants, trees and minerals, to the virtues and uses of which we are yet strangers. What the soil is capable of producing can only be guessed at and known by experiment. Reasoning from Analogy we may suppose that all the rich productions of Asia may in time be transplanted hither. Agriculture is in its infancy. The human mind seems just awakening from a long stupor of many ages to the discovery of useful Arts and inventions. Our governments are yet unformed and capable of great improvements in police, finance and commerce. The history, manners and customs of the Aborigines are but little known. These and a thousand other subjects which will readily suggest themselves open an inexhaustible mine to men of a contemplative and philosophical turn."[1]

Above all, Jefferson's America was the exact opposite of the climatically and intellectually cold and enfeebling place portrayed by Buffon—a man who had never visited any part of the New World. Americans themselves were becoming a people to be reckoned with. No longer a stepchild of Europe, America was emerging as a world force. It had fought successfully with Britain against France in the French and Indian War and against Britain in the American Revolution. Now the population was growing by leaps and bounds.

On the central issue—the supposed inferiority of American wildlife and indigenous peoples—Buffon's slurs had to be answered, not just because he had expressly stated that nature in the Americas was inferior and either primitive or degenerate but because he also claimed to identify the causes of that inferiority in the basic inadequacy and degeneracy of the land itself. Other European authors had already copied and even embellished Buffon's slanders, broadcasting them to wider and wider audiences. Rebuttal was necessary, not least because Franklin and Adams were just then in Europe trying to negotiate loans to support the new nation.

In composing *Notes,* Jefferson not only expanded on his answers to Barbé-Marbois's questions about Virginia but reordered the questions. Instead of beginning with the dry data about charters and constitutions that Barbé-Marbois wanted, Jefferson introduced the land itself—Virginia and America—in his first five chapters. He laid out what soon became recognized as the first authoritative geography of the Unites States and much of the land extending to the west—the "Louisiana Country" that controlled the navigation of the Mississippi and Missouri Rivers, territory that Jefferson coveted as essential for the future of the republic. The structure of his initial chapters corresponds to Barbé-Marbois's inquiries about rivers, sea ports, mountains, cascades, mines, and minerals.

And all are oriented toward travel and commerce. In addition to describing the major rivers of Virginia, for example, Jefferson used information from his many correspondents to describe the Mississippi, which "will become one of the principal channels of future commerce for the country westward of the Alleghany," Missouri, Illinois, Kaskaskia, and Ohio Rivers, plus all the other major rivers of the eastern part of the continent, including the "Tanisee" (Tennessee), Cumberland, Wabash, Kentucky, and the Great Kanhaway Rivers. To all of this Jefferson added practical information. "The country watered by the Missisipi and its eastern branches, constitutes five-eights of the United States," but "navigation through the Gulph of Mexico is so dangerous, and that up the Mississipi so difficult and tedius, that "there . . . will be competition between the Hudson and Patowmac Rivers for the residue of the commerce of all the country westward of Lake Erie."

With his constant goal of promoting the glories of America, he could not resist adding a page on what he considered the "most sublime of

Nature's works," the remarkable "Natural Bridge" near Roanoke. Jefferson loved this place so much that he bought it solely to preserve it; this was surely the first "national park" in America. "It is impossible for the emotions arising from the sublime, to be felt beyond what they are here: so beautiful an arch, so elevated, so light, and springing as it were up to heaven, the rapture of the spectator is really indescribable!"

Notes on the State of Virginia is perhaps most famous today for its discussion of American natural history, although Barbé-Marbois had not requested information on natural history per se. In that area, focusing on economic issues, he had asked only for an accounting of "Productions Trees Plants Fruits and other natural Riches."

Jefferson's section "Vegetables" in *Notes* appears to have been duplicated from his report to Barbé-Marbois. He listed 138 kinds of native American plants, most as individual species, all with their common names and formal scientific names, limiting himself to useful kinds but noting that "there is an infinitude of other plants and flowers." His lists were grouped into four categories. Among those "useful for medicine" were senna, Jimson weed, various mallows, pleurisy root, snakeroots, valerian, gentian, ginseng angelica, and cassava. Jefferson added a footnote on the halucinogenic properties of the poisonous Jimson weed ("James-town weed"), *Datura stramonium.* The list of "Esculents" (food plants) included artichokes, potatoes, a variety of grains (oats, Indian millet, others), hops, cherry and other fruit-bearing trees, such as crabapple, mulberry, walnut, chestnut, and hazelnut, and soft fruits, such as grapes, strawberries, whortleberries, gooseberries, cranberries, raspberries, and cloudberries. Among the trees, Jefferson listed one of his favorites, the "Paccan," being the first to record what botanists had missed: that it was a different species from the white walnut.

Among the "ornamental" species listed were familiar trees like the "popular," black poplar, aspen, catalpa, magnolia, sassafras, locust, dogwood, redbud, barberry, holly, elder, "papaw," honeysuckles, and "sumach." Finally, for "fabrication" he listed hemp, flax, pines, hemlock, oaks, birches, elm, willow, and sweet gum, among others. Jefferson listed separately the crop plants "handed along the continent from one nation to another of the savages": tobacco, maize, round potatoes, pumpkins, "Cymlings and Squashes" (the last three all being species of *Cucurbita*).

Jefferson turned next to animals. This section constituted the meat of his defense against Buffon. Jefferson knew that with his superior knowledge of American wildlife he could take on Buffon directly. Having first attacked him on the subject of the incognitum (the mastodon), he proceeded to refute Buffon's claims about (living) American mammals. "The opinion advanced by Count de Buffon is 1. That the animals common both to the Old and New World, are small in the latter. 2. That those peculiar to the new are on a smaller scale. 3. That those which have been domesticated in both, have degenerated in America: and 4. That on the whole it exhibits fewer species. And the reason he thinks is, that the heats of America are less; that more waters are spread over its surface by nature, and fewer drained off by the hand of man. In other words, that heat is friendly, and moisture adverse to the production and development of large quadrupeds."

In reply, Jefferson pointed out that no data supported a theory that the American climate was wetter. Nor, in any case, was the state of knowledge of animal husbandry sufficient to allow the conclusion that moisture was "unfriendly to animal growth." Jefferson neatly turned the tables by pointing out that Buffon himself had *denied* that cold was injurious to large mammals. Buffon had written, "In general it seems that somewhat cold countries are better suited to our oxen than hot countries, and they are the heavier and bigger in proportion as the climate is damper and more abounding in pasture lands."

In answering Buffon in detail, Jefferson's first strategy was to reduce the scope of the comparison. Buffon had treated "America" as if it were one place, and he had compared the entire New World with the Old World. He had stated, inaccurately, that the largest mammal in the New World was the tapir, which was nothing like as large as the Old World elephants, giraffe, hippopotamus, or rhinoceros. (In fact, in his lists, the bison at 1,800 pounds was the biggest American mammal.) Jefferson controlled the argument by limiting his comparisons to North America and Europe. This was a clever lawyer's device, amply justified by the focus of his book on "Virginia." It allowed him to point out that although the tapir was perhaps smaller than the largest Old World mammals, it was larger than any *European* mammal. "To preserve the comparison, I will add that the wild boar, the elephant of Europe, is little more than half that size."[2]

Instead of arguing qualitatively, Jefferson made his case with data abstracted from standard European reference works and collected from his own observations and from information collected from a wide range of friends.[3] One of the areas in which he requested help concerned the problem of distinguishing among the various animals called round-horned elk, palmated or flat-horned elk, black moose, gray moose, and orignal.[4] The questions that Jefferson asked were detailed, down to "Do their feet make a loud rattling as they run." The letters show that Jefferson had completed a draft of *Notes* by mid-1783 and had sent it out to friends for comment but was still adding information to his text as late as April 1784, just before he left for Paris. To this particular set of questions, the answers boiled down to an agreement that only three species were involved: wapiti or elk (round-horned), moose (flat-horned), and caribou (reindeer).

Looking at things like a lawyer, Jefferson did not have to prove that mammals in America were bigger than in Europe, just that they were not inferior. Jefferson made his case on the evidence, laid out in three sets of tables of comparison. First, having pointed out that the tapir was bigger than any European mammal and shown that the mastodon was an animal bigger than either the European mammoth or the African and Asian elephants, Jefferson tabulated evidence on the twenty-six mammal species then thought to be common to Europe and North America. Of these, "7 are said to be larger in America, 7 of equal size, and 12 not sufficiently examined. So that the first table impeaches the first member of the assertion, that of the animals common to both countries, the American are smallest, 'et cela sans aucune exception' [translation: "and that without a single exception"]. It shews it not just, in all the latitude in which its author has advanced it, and probably not to such a degree as to found a distinction between the two countries."

As Jefferson knew, there were too few data to make a firm case. "There remain then the buffalo, red deer, fallow deer, wolf, roe, glutton [wolverine], wild cat, monax [groundhog], vison [bison], hedge-hog, martin, and water rat, of the comparative sizes of which we have not sufficient testimony. It does not appear that Messrs. de Buffon and D'Aubenton have measured, weighed, or seen those of America." So, even-handedly (and advantageously) he concluded that the available results "justify a suspension of opinion until we are better informed, and a suspicion in the mean time that there is no uniform difference in favour of either."

Publication of the *Notes* produced more information. For example, Chastellux and others had realized that the American (white-tailed) deer was not the same as the European deer. And James Madison wrote on May 12, 1786, pointing out some errors in the first Paris edition. "I observe that in your Notes you number the fallow and Roe-deer among the native quadrupeds of America. As Buffon had admitted the fact, it was, whether true or erroneous, a good argument no doubt against him. But I am persuaded they are not natives of the New Continent."[5] Jefferson, who told George Wythe that he was revising the book for printing in London (the Stockdale edition of 1787), did not make the change, although he eventually realized that the large mammals of North America, although filling similar ecological niches, were all different species from those in Europe.

In the text accompanying the tables of mammals, Jefferson dissected what Buffon had written with a deft verbal scalpel. Nor was he beyond yet another rhetorical dig. "One sentence of his book must do him immortal honour. 'I love as much a person who corrects me in an error as another who teaches me a truth, because in effect an error corrected is a truth.'" Jefferson had no doubt how little Buffon would enjoy being corrected.

Jefferson's second table, of mammals that were indigenous only to one or the other continent, showed that the North American species were not only larger but more numerous (by 74 to 18). Here again Jefferson was helped by being able to include species from the Central American region in his list.

Finally, Jefferson assembled a table showing that European animals introduced to the New World were not smaller than the originals. This was his least well-supported claim. It is not clear why Jefferson gave a maximum weight for the horse, hog, sheep, and goat in America but gave no comparative weights for the European species, nor why he listed a maximum for

North America cattle of 2,500 pounds, compared with only 763 pounds for European cattle. As an agriculturalist, Jefferson had easy access to data that would have shown him higher weights for cattle in Europe than that. The weight of cattle sent to market in England doubled in the seventeenth century; in 1692 a prize ox in Lincolnshire tipped the scales at 3,564 pounds.[6]

The source of Buffon's charge that European species degenerated when they were imported to America seems to have been the Swedish-Finnish botanist Peter Kalm. In his 1748 book of travels, Kalm had noted that at Philadelphia "the cattle degenerate by degrees here, and become smaller."[7] But Jefferson knew the reason: American cattle were smaller because the climate was so equable that they could be left out all winter to forage. If they had been brought into barns for the winter and fed hay, like European cattle, they would have grown much larger. The same point had been made a fifty years earlier by William Byrd.[8]

Jefferson sent Buffon a copy of his book (via Chastellux), and when the two men finally met, Buffon airily dismissed Jefferson's work. As Jefferson later recounted: "I was introduced to him as Mr. Jefferson, who, in some notes on Virginia, had combated some of his opinions. Instead of entering into an argument, he took down his last work, presented it to me, and said, 'When Mr. Jefferson shall have read this, he will be perfectly satisfied that I am right.'"[9] It was at this point that Jefferson decided to get hold of a moose skeleton to demonstrate to Buffon that there was a mammal in America bigger than anything in Europe.[10] Together with the moose and other materials, Jefferson sent a covering letter that demonstrated, inadvertently, that confusion still existed over the names "elk" and "moose": "I really doubt whether the flat-horned elk exists in America: and I think this may be properly classed with the elk, the principal difference being in the horns. I have seen the Daim, the Cerf, the Chevreuil of Europe. But the animal we call Elk, and which may be distinguished as the Round-horned elk, is very different from them. I have never seen the Brand-hirtz or Cerf d'Ardennes, nor the European elk. Could I get a sight of them I think I should be able to say to which of them the American elk resembles most, as I am tolerably well acquainted with that animal . . . I really suspect you will find that the

Moose, the Round horned elk, and the American deer are species not existing in Europe. The Moose is perhaps of a new class."[11]

Jefferson's list of the birds of the New World was largely taken from Mark Catesby's explorations in the Carolinas and the West Indies between 1714 and 1719.[12] But from his own experience Jefferson was able to add to that list of seventy-six the names of thirty-two species that Catesby had not seen. Jefferson seems to have been personally familiar with eighty to a hundred species of North America birds.[13]

A note must be added here about Jefferson's use of the most modern scientific approach to naming species. For the birds and the plants, Jefferson provided both the common names and their formal scientific equivalents in the new binomial system invented by Linnaeus (the system that famously defines humans as *Homo sapiens*). His tables of mammals used only the common names—perhaps because of the confusion over specific identities. Jefferson believed that using the formal Latin names would make the lists more understandable to foreigners than using the popular (common) names in English. No doubt he was also showing off when he gave a description of the "Paccan" tree (pecan), which was not listed by Linnaeus, in the old classical style: "foliolis lanceolatis, acuminates, serratis, tomentosis, fructu minore, ovato, compresso, vix insculpto, ducli, utamine, tenerrimo."[14]

In using the Linnaean system of names and classifications, Jefferson showed himself in step with modern European thinking and far ahead of most of his American contemporaries.[15] As he wrote to a friend much later, Linnaeus's system was "adopted by all, and united all in a common language. It offered the three great desiderata . . . of aiding the memory," "rallying all to the same names for the same objects," and "enabling . . . when a subject was first presented, to trace it by its character up to the conventional name by which it was agreed to be called."[16] Jefferson had high hopes that new systems of nomenclature would similarly transform two other fields: mineralogy and chemistry. But the advantages and disadvantages of their competing systems were as yet difficult to adjudicate.

A minor puzzle needs to be solved here. It has been stated that Jefferson acquired his copies of Linnaeus's work on March 3, 1785, when he purchased them from the library of a friend, the Reverend Samuel

Henley.[17] If Jefferson had bought the books in 1785 (when he was already minister in Paris), he would not have had them in time to consult them for the Latin names used in *Notes.*

Long after Jefferson's mentor William Small had left the College of William and Mary, Henley had been professor of moral philosophy there. A clergyman originally from England, he returned to England in 1775, ahead of the conflict, leaving his library behind under the care of the Reverend James Madison, then president of the college. His instructions were that Madison should, as Jefferson said, "dispose of [the books] generally, or to let me have such as I wished to possess." Madison showed them all to Jefferson and set a price for the ones that Jefferson wanted.

In fact, Jefferson had the books in his possession seven years earlier.[18] March 3, 1785, was simply the date when Jefferson, following up on previous letters and lists that had gone astray, sent to Henley from Paris a list of books he had taken from the collection.[19] In one of those earlier letters, written June 9, 1778, he told Henley that he had chosen the books he wanted and offered to buy them "at sterling cost and (I) would remit you the money by way of France."[20] He had plenty of time, therefore, to annotate his plant lists in *Notes* with the correct Linnaean names.

Jefferson was just as opposed to Buffon's natural philosophy and theories as he was to his mangling of the facts of American natural history. It is possible that Buffon had a political animus when he asserted in *Histoire Naturelle* that all of North America was cold, wet, and unpropitious for the development of healthy flora and fauna or for human existence, but he wrote, or claimed to write, on the basis of a scientific thesis. He thought that the differences between the New and the Old Worlds were explained by climate and by geographical and geological conditions, and he developed these ideas in his ninth volume of 1761 and the fifth supplementary volume of 1778. Buffon did have some evidence on his side. As he pointed out, cities on the same latitude—Quebec and Paris, for example, and Boston and Seville—had very different climates, the North American ones always being colder. Edinburgh, a cool but nonetheless inhabitable city, lies on the same latitude as Hudson Bay.

In his early volumes, Buffon proposed a variety of explanations for the supposed cold and damp of "America." The inferiority of the New World, he wrote in volume 9, "must be referred to the quality of the earth and atmosphere, to the degree of heat and moisture, to the situation and height of mountains, to the quantity of running and stagnant waters, to the extent of forests, and, above all, to the inert condition of Nature in that country. In this part of the globe, the heat in general is much less, and the humidity much greater . . . The east wind, which blows perpetually between the Tropics, arrives not in America, till it has traversed a vast ocean, by which it is cooled. Hence this wind is much cooler in Brasil, Cayenne, &c, than at Senegal, Guinea, &c, where it arrives impregnated with the accumulated heat from all the lands and brining sands on its passage through Asia and Africa . . .

"This first cause renders all the east coast of America much more temperate than Africa or Asia." The wind then gets hotter as it goes over land until it reaches the Andes, where it is "stopped and cooled . . . Besides, as the earth is every where covered with trees, shrubs, and gross herbage, it never dries . . . In these melancholy regions, Nature remains concealed under her old garments . . . being neither cherished nor cultivated by man . . . The scarcity of men, therefore, in America, and most of them living like brutes, is the chief cause why the earth remains in a frigid state."[21]

From all this, Buffon concluded that the New World, except for its obviously ancient far-western mountains, was "more recent, and has continued longer than the rest of the globe under the waters of the ocean." It really was a new world. The evidence for this was partly geological. In the "new lands, elevated and formed by the sediments of waters . . . in many places, immediately under the vegetable stratum, we find large masses of limestone, but which are softer than our free-stone."

Buffon supported his theory by arguing that "if this continent is really as ancient as the other, why was it so thinly peopled?" And why are the people "still ignorant of the art of transmitting facts to posterity by permanent signs? Their arts, like their society, were in embryo; their talents were imperfect, their ideas locked up, their organs rude, and their languages barbarous."[22] He theorized that an important cause was the primitive state of Native Americans' culture. Their basic lack of numbers, limited agriculture, and use of primitive tools meant that they had not cleared the forests, drained the swamps, or planted crops and had thus not caused the climate to be ameliorated. In heavily populated Europe, on the other hand, millennia of agriculture had produced lands more suitable for both wildlife and people.

In the supplementary volume of 1778, Buffon published his highly celebrated essay "Époques de la Nature," in which he expanded on his long-held view, shared with René Descartes, that the earth had once been a molten ball spun off from the sun. Buffon elaborated his idea of a cooling earth to define seven stages, or epochs, of prehistory. (The analogy with the seven days of biblical Creation was not accidental.)

In Buffon's first epoch "the earth and planets assumed their proper form." In the second epoch, the fluid earth cooled enough to consolidate. In the third epoch, the continents were covered with water and life began. The waters receded in the fourth epoch, giving dry land, and this was when volcanoes first erupted. In the fifth epoch "the elephants, and other animals of the south, inhabited the northern regions. In the sixth epoch the continents separated and the seventh epoch was marked by the arrival of humankind and its influence on nature.

The key stage was Buffon's fifth epoch. In Buffon's theory, the earth had cooled first at the poles, where the thickness of the insulating crust of the earth was least. As the proto-earth cooled, it was therefore at the poles that life could first arise. The new life forms gradually spread southward with the continued cooling of the earth and were followed by waves of more species. In Buffon's fifth epoch conditions at the North Pole were still similar to those currently enjoyed by the tropics and were perfectly suited to the origin of large land mammals, including the elephant, the rhinoceros, and the hippopotamus.

In the sixth epoch, after the continents had separated, the continued cooling of the Arctic regions led to extinctions and migrations. In the Old World, the direct descendants of the large early mammals moved to occupy more southerly regions. "In the progress of time, the elephant gradually migrated to the climates under the torrid zone, which alone have longest preserved, by the greater thickness of the globe, a superior degree of internal heat." But the passage to warmer climates in the New World was supposedly blocked by mountains. Therefore the New World, once the original elephants and other large land animals had been extinguished by the cold winters, was inhabited only by smaller, lesser, creatures. In just a few pages of *Histoire Naturelle,* Buffon had reduced the wildlife of the entire New World to secondary status in the sweeping history of the earth and had declared its mightiest denizen—the mastodon—dead.

Jefferson believed that the earth had been created by God as described in the Book of Genesis, and he showed that Buffon's theory, at least as in connection with the origins of American wildlife, could be invalidated. Religion aside, he saw three obvious flaws in Buffon's theory. First, he argued that if indeed the earth has been cooling, then its inner heat would be more insulated from heat loss at the tropical regions, where the earth's crust was thicker, and the heat should have been more easily lost at the poles, where the crust was thinner. His second objection was entirely factual: there was no great mountain barrier to prevent the southward migration of large mammals. A *sea* barrier between North and South America, Jefferson noted, it "would have served that purpose" if Buffon had proposed one. A third, even more basic objection was that the mastodon might not be extinct.

The only alternative to either Buffon's ingenious theory or to the Creation story seemed to be a catastrophe of some sort—that after the earth had been created (more or less in its present state), there had been some upheaval or revolution causing the formerly warm climates of the northern latitudes suddenly to become cold. The most likely cause would have been a change in the earth's rotation around its axis. For Jefferson, such an event was too complicated and improbable to be an acceptable explanation. The simplest logical explanation was that the world had been made as it is now and that different species had been created for different parts of both the New and Old Worlds, each with its own geographical (latitudinal) ranges. The mastodon was the proof of this. But accepting this theory left all sorts of other problems unresolved.

CHAPTER SEVEN

Mountains and Shells

IN 1769 THE SCIENTIFIC WORLD was agog at the possibilities presented by a transit of Venus. Venus would pass directly in front of the sun's disk, and sightings taken from different points on the earth's surface would, by triangulation, allow astronomers to produce an estimate of the distance between the earth and sun and to help refine the dimensions of the whole solar system. Captain James Cook sailed to Tahiti on HMS *Endeavour* to observe the event. It could also be seen across the continent of North America, and as a dramatic demonstration of the new laws of cosmology derived by Copernicus, Kepler, and Newton, it gave special impetus to natural philosophy there.

The very first issue of the *Transactions of the American Philosophical Society* contained no fewer than nine papers by American observers of the 1769 transit. That volume also contained a paper with a recipe for making red currant wine. Similarly, when the American Academy of Arts and Sciences was founded in Boston by John Adams in 1780 (as a rival to Philadelphia's American Philosophical Society), the first issue of its *Memoirs* contained thirteen papers on astronomical subjects, including a solar eclipse and a transit of Mercury. There was also a letter discussing whether swallows, instead of flying south in winter, hibernated in mud at the bottom of ponds.[1]

The juxtaposition of these essays in scholarly journals might, at first glance, seem anomalous; but it is not.

Science at any moment is a mixture: the old is being replaced by the new. Scientific knowledge is a mélange of long-held understandings, even folk wisdom, and new observational and experimental discoveries. At every stage since the authority of Aristotle, Pliny, Ptolemy, and the old church was abandoned, science has constantly been on the cusp of change, between ancient and modern. There never has been a time of perfect stability in the world of science; whenever new, apparently immutable truths are laid down as signposts for our philosophies, hard-won positions have to be abandoned. As Jefferson had said, "Science is progressive . . . What was useful two centuries ago is now become useless . . . What is now deemed useful will in some of its parts become useless in another century."

Jefferson's writings on science, as presented in *Notes on the State of Virginia* and his lifelong correspondence, demonstrate the changing state of knowledge in the eighteenth and early nineteenth centuries, and also the changing nature of science itself. Science was steadily becoming not one subject but a whole series of separate disciplines, like physics, chemistry, and geology. The whole of science was loosely held together by a general methodology involving hypothesis, observation, experiment, and analysis, but each "field" was acquiring its own methods and principles and, in many cases, connections to technology and invention.

A great deal of science in Jefferson's time was on the cusp of change between traditions dependent on the biblical account of Creation or inherited from the ancient Greek philosophers and those belonging to the modern Age of Reason. Thus Jefferson's analysis of the mastodon had been nearly impeccable, and his understanding of steam power was advanced, whereas Buffon (and the Reverend Ezra Stiles at Yale) still gave partial credence to the existence of human "giants." When it came, however, to the vast and unknown American West—the lands toward the Great Stoney Mountains (the Rocky Mountains)—Jefferson used the evidence of travelers' tales to support his conviction that the mastodon was not extinct. He was foremost among his contemporaries in his admiration and understanding of Newton's mechanics and mathematics, including the calculus, but he always denied, on grounds of biblical authority, that extinction of species ever occurred.

Religion affected the scientific thinking of Jefferson and his contemporaries in two ways. Jefferson, for all his doubts about conventional revealed religion, believed in a God who was the creator of all. And thus he gave credence to the account of Creation in Genesis and its implication that the earth was very young and that life on earth had been created only once. He may have had his doubts about Noah's Flood, or at least its scope, but he seems to have believed in God's promise, when he gave the earth back to Noah's descendants after the Flood, "I will not again curse the ground any more," which meant that the earth could not have changed since that date.[2]

Many European natural philosophers of the late eighteenth century concluded, on the other hand, that Genesis was a series of fables; that the earth had spun off from the sun tens or hundreds of thousands of years ago, as a still-molten ball and whirled through space into its present orbit around the mother star. Buffon theorized that the sun had been hit by a huge comet that blasted off the earth's portion. To anyone who took the first chapter of Genesis as the literal truth, this was blasphemous nonsense.

Among the other contentious issues of late eighteenth-century science was the interpretation of meteorites: If they were "stones falling from the sky" where did they come from? And how old was the earth? Was there really a Noah's Flood? How are weather patterns caused and controlled? What was the explanation for the Gulf Stream? What were the causes and meanings of the fact that humans exist in the form of several superficially quite different races? Has life on earth changed over time? Have any creatures become extinct? How could longitude easily be calculated? What were double stars? What were comets?

The list could be endless. There were also the practical questions. Could steam be harnessed to power a carriage or a boat? Could a vessel be built to move underwater? Could humans find a way to fly in the air? Does human activity change climates and weather patterns? Most of these questions had been addressed by a long line of thinkers, including Leonardo da Vinci and, more recently, French and English philosophers. Many of them resonate with us today because, in new forms, they are still being posed. They are all ones to which Jefferson gave a great deal of thought, including a final question being asked then, just as it is now: Was science and technological progress a good thing?

One of the most difficult subjects in which Jefferson sought to find philosophical certainty was geology. Of the sciences undergoing revolutionary changes during Jefferson's lifetime, geology most disturbed conventional views of biblical truth and even common sense.[3] For millennia, most people had taken the earth as a metaphor for all that is unchanging. The earth with its eternal hills was solid as a rock. Its rivers had flowed forever in the same courses. Forests might be cut down and fields planted, but the underlying landscapes were always the same. It had all been created in perfect form for human use, and if anything major had perchance changed, then that must have happened with the great flood of Noah. Furthermore, because the earth seemed unchanged and unchanging, there was no reason to doubt the Judeo-Christian tradition that it was only a few thousand years old.

Three approaches to science upset this view. The first was the most obvious: observation. When men ventured beyond their green fields and valleys, where the deeper structure of the rocks was covered over by grass and trees, and lifted their eyes "unto the hills," and when they looked with unbiased eyes, they saw irregularities. The earth might seem stable now, but at some time in the past it had been subject to violent forces. As Thomas Burnet wrote in 1681, after traveling in the Swiss Alps: "We must . . . be impartial where the Truth requires it, and describe the Earth as it is really in its self; . . . 'tis a broken and confus'd heap of bodies, plac'd in no order to one another, nor with any correspondency or regularity of parts: And such a body as the Moon appears to us, when 'tis look'd upon with a good Glass, rude and ragged . . . a World lying in its rubbish."[4]

Curiosity was another powerful impetus for change in science. The rocks making the earth's crust were not only broken and thrown about; some also appeared to have been folded and faulted, others, raised and lowered, presenting a picture of once-great fluidity now frozen in time. How had that been caused? Nor was the earth entirely stable at present either; it was shaken by earthquakes and volcanoes, evidence also that under the surface there lay a roiling molten core. What else was underneath the solid crust of the earth? Was it all molten? Were there vast caverns, as some thought?

If the earth's crust had been dramatically changed, mechanisms had not only to be thought about but to be made consistent with the Bible. Most theoretical approaches involved one or more great "catastrophes"—

perhaps nothing more complicated than a series of floods. The Bible itself gave some clues here. At Noah's Flood, the water came from both the forty days and nights of rain and a venting of the "fountains of the deep." What else could have caused changes?

A third impetus for change, therefore, was theory. One issue that focused people's minds on geology was the age-old problem that one could find, high up on mountains, even the lowly Blue Ridge Mountains of Virginia, beds of petrified seashells. That these were real shells that had once housed living creatures was demonstrated by careful observation. The shells were preserved in different growth stages and in lifelike positions; some were found clinging to each other. Did these shells belong to the same species that currently lived on the seashore? That was hard to say, but most authors said yes. Burning the shells often gave rise to a smell of burning horn. It was tempting to conclude that the deposits of petrified shells, together with fish teeth and other fossils, were direct evidence of the awful mortality produced by the Noah's Flood. But if the animals had all been killed by a flood, their remains should in theory have been washed back into the sea or deposited along the coasts, not stranded on mountains.[5] If, on the other hand, those sea creatures had lived in the sediments where they were preserved, then somehow the mountains themselves had been raised up as the result of stupendous changes in the surface of the earth, post-Creation.

Many observers noted that an insidious force—erosion—constantly wears away the high land. Was it possible that the earth would eventually become entirely flat? Indeed, had it originally been made as a flat as an egg, before God raised up the mountains? Scholars like Burnet proposed "catastrophe theories" in which the present appearance of the earth was the result of a single cataclysmic event involving the Flood. There was also the possibility that the earth had been impacted by a comet, or more than one; comets had become the object of general fascination after Edmund Halley in 1705 had worked out that the comet that now bears his name had a regular periodicity, passing within sight of earth every seventy-five or seventy-six years.[6]

Nicolas Steno (Niels Stensen), a Danish anatomist and geologist working at the Accademia del Cimento (Academy of Experiments) at Florence in 1669, proposed a theory that the earth had been changed by a sequence of revolutions. Periods of erosion and deposition of strata were followed by subsidence of the crust into vast underground caverns.[7] One of the more persuasive aspects of this theory was that it did not require that mountains be raised up from the sea; they were merely left in place when the earth all around collapsed. The sea creatures that had lived in the ancient ocean were then stranded high and dry. (Steno's theory had just enough congruence with the Bible to get past the censors of the Inquisition.)

At almost the same time, the English scientist Robert Hooke suggested that the earth was constantly being remodeled by a combination of offsetting factors. Erosion carried sediment down to the sea to be deposited and consolidated, eventually forming new strata of rock. These were progressively modified by the earth's inner heat and elevated by volcanoes and earthquakes. The hills and mountains so raised were then subject to erosion again, and all was recycled over and over.[8]

A major problem with any theory of geomorphology was that no one could imagine how such dramatic changes as the removal of old mountains and the creation of the Alps and Andes (or the evident folding of the Blue Ridge Mountains of Virginia) could have been accomplished in the time span conventionally allotted to the age of the earth—six thousand years. Sober observers concluded that the earth had to be much older than could be construed by reading the Bible.

Jefferson did not believe that the earth was totally unchanged since Creation. In his chapter on mountains in *Notes,* he used his knowledge of the map of Virginia by Joshua Fry and Peter Jefferson and observations made during his own extensive travels to record that the Appalachian Mountains were arranged as a series of southeast-northwest parallel ridges broken through at various places by rivers running at right angles to the ocean. These gaps had evidently been formed after the ridges were put in place. He envisaged the Shenandoah River having "ranged along the foot of the mountains an hundred miles to seek a vent." Then it met the "Patowmac," and "in the moment of their junction they rush together against the mountain, rend it asunder, and pass off to the sea . . . The piles of rock on each hand . . . are evident marks of their disrupture and avulsion from their beds by the most powerful agents of nature." Our "first glance" thus "hurries our senses into the opinion, that *this earth had been created in time,* that the mountains were formed first, that the rivers began to flow afterwards" (emphasis added). The italicized words later got Jefferson into trouble with biblical literalists, for whom the earth was not created "in time" or "over time" but instantly.

The problem of petrified seashells had the potential to unlock the mystery of the origin of mountains, and this became a challenge to Jefferson. Not only was he familiar with all the literature on the subject, he had observed for himself that "near the eastern foot of the North mountain are immense bodies of *Schist,* containing impressions of shells in a variety of forms. I have received petrified shells of very different kinds from the first sources of the Kentucky, which bear no resemblance to any I have ever seen on the tide-waters. It is said that shells are to be found in the Andes, in South America, fifteen thousand feet above the level of the ocean."[9]

As he considered rival geological theories in *Notes,* Jefferson accepted that the petrified shells were the remains of real once-living organisms. But he rejected the possibility that they were evidence of a great deluge. He used simple calculations to show that no flood of water could ever have been mighty enough to account for the location of seashells on mountaintops (especially the shells in the high Andes). But if tides had not risen to the height of the mountains, then logically a "second opinion has been entertained, which is, that, in times anterior to the records either of history or tradition, the bed of the ocean, the principal residence of the shelled tribe, has, by some great convulsion of nature, been heaved up to the heights at which we now find shells and other remains of marine animals."

A third solution, for which Jefferson's immediate source was Voltaire, although it was much older, was that the shells were not real shells after all but were artifacts. Somehow "nature formed these apparent shells directly in the rocks" and was able to inject "the calcareous juice into the form of a shell." When in France, Jefferson took time during a trip to southern France and Italy to check on a story that had been gaining currency: that shell-like structures could be observed actually growing in the rocks.

To his great dissatisfaction, Jefferson could not resolve the matter of mountains and shells, dryly remarking in *Notes,* "There is a wonder here somewhere." Finding none of the three solutions to the puzzle satisfactory, he wrote, "Ignorance is preferable to error, and he is less remote from the truth who believes nothing, than he who believes what is wrong." This did not mean, however, that he lost his interest in the matter of shells and mountains. And petrified shells were, as the Reverend James Madison soon reminded him, also found on coastal plains.[10]

Two years after the first printing of *Notes,* in a letter to David Rittenhouse in Philadelphia, he summed up his thinking again. Now he saw four possibilities to account for fossil shells being found on mountains. "1. That they have been deposited there by even in the highest mountains by a universal deluge. 2. That they with all the calcareous stones and earths are animal remains. 3. That they grow or shoot as chrystals do . . . Another opinion might be added . . . , that some throw of nature has forced up parts which had been the bed of the ocean. But have we any better proof of such an effort of nature than of her shooting a lapidific juice into the form of a shell?"[11] Once again, Jefferson found himself in a position that he could not resolve either by logic or by science. It is fascinating that the state of science as Jefferson knew it made it just as likely—and improbable—that fossils grew by themselves within rocks as that mountains had been raised from the seabed. Everyone knew that rocks came in various odd shapes, like hearts and shoes, which seemed to have nothing to do with real, once-living animals.[12] No one had seen a whole mountain move, nor was there any biblical authority for such an event.

Jefferson was right that "there is a wonder here." No one had a satisfactory geological explanation that fit with the available observational data. Steno's scheme supposed upheavals in the earth's crust on an unimaginable scale. Hooke's scheme invoked earthquakes, but there were many mountains without seismic or volcanic activity. For all Jefferson's interest in geology, and particularly in the rocks themselves, he had no vehicle to link his naturalist's enthusiasm with his philosopher's search for causes and laws.

Shortly after Jefferson completed the manuscript of *Notes,* his friend

Charles Thomson wrote to him suggesting that he read a new theory by the English geologist John Whitehurst, and Jefferson at once bought a copy of Whitehurst's *An Inquiry into the Original State and Formation of the Earth Deduced from Facts and the Laws of Nature* (1778).[13] Whitehurst's theory was that the earth began as a whirling fluid "chaos" that naturally assumed an oblate spheroidal shape as it rotated. As it slowed down, the different components—air, earth, and water—separated out, and the various solid parts were deposited according to their weight, creating the strata. A variety of shellfish and bony fish were created to live in the ancient seas, and life appeared on one or more islands. Whitehurst's new wrinkle on the "catastrophic flood" theory proposed that at the biblical Flood, water penetrating deep into the earth became superheated by subterraneous fires. "Two oceans of melted matter and water came into contact, whence a violent explosion ensued, which tore the globe into millions of fragments . . . [some] being more elevated, and others more depressed." This explosion threw up mountains and continents and thus elevated strata (with shells) that had once been at the bottom of the sea.

Being resident in Paris, Jefferson was aware as never before of the potential of steam power. "The power of this agent, tho' long known, is but now beginning to be applied to the various purposes of which it is susceptible." But he rejected Whitehurst's theory in a memorable letter that formed one of his minor scientific dissertations and shows us the pattern of his scientific thinking. Jefferson's first resort was rhetorical (even sarcastic); its logic depended on a belief in the Creator. "You observe that Whitehurst supposes [steam] to have been the agent which, bursting the earth, threw it up into mountains and vallies . . . There are great chasms in his facts, and consequently in his reasoning. These he fills up by suppositions which may be as reasonably denied as granted. A sceptical reader therefore, like myself, is left in the lurch.

"I acknolege however he makes more use of fact than any other writer of a theory of the earth. But I give one answer to all these theorists. That is as follows: they all suppose the earth a created existence. They must suppose a creator then; and that he possesed power and wisdom to a great degree. As he intended the earth for the habitation of animals and vegetables is it reasonable to suppose he made two jobs of his creation? That he first made a chaotic lump and set it into rotatory motion, and then waiting the millions of ages necessary to form itself, that when it had done this he stepped in a second time to create the animals and plants which were to inhabit it? As the hand of a creator is to be called in, it may as well be called in at one stage of the process as another. We may as well suppose he created the earth at once nearly in the state in which we see it, fit for the preservation of the beings he placed on it."[14]

As for the theory that the earth had been a rotating, molten sphere that was forced into an oblate spheroidal shape, slightly wider at the equator, he neatly turned the argument on its head. An oblate spheroid was simply the shape that God would have chosen in the first place. "I suppose that the same equilibrium between gravity and centrifugal force which would determine a fluid mass into the form of an oblate spheroid, would determine the wise creator of that mass, if he made it in a solid state, to give it the same spheroidical form. A revolving fluid will continue to change it's shape till it attains that in which it's principles of contrary motion are balanced; for if you suppose them not balanced, it will change it's form. Now the same balanced form is necessary for the preservation of a revolving solid. The creator therefore of a revolving solid would make it an oblate spheroid, that figure alone admitting a perfect equilibrium. He would make it in that form for another reason, that is, to prevent a shifting of the axis of rotation.

"Had he created the earth perfectly spherical, it's axis might have been perpetually shifting by the influence of other bodies of the system, and by placing the inhabitants of the earth successively under it's poles, it might have been depopulated: whereas being Spheroidical it has but one axis on which it can revolve in equilibrio . . . We may therefore conclude it impossible for the poles of the earth to shift, if it was made spheroidically, and that it would be made spheroidal, tho' solid, to obtain this end. I use this reasoning only on the supposition that the earth has had a beginning. I am sure I shall read your conjectures on this subject with great pleasure."

And, "With respect to the inclination of the strata of rocks, I had observed them between the Blue ridge and North Mountain in Virginia to be parallel with the pole of the earth. I observed the same thing in most instances in the Alps between Nice and Turin: but in returning along the precipices of the Appennines where they hang over the Mediterranean, their direction was totally different and various; and you mention that in our Western country they are horizontal. This variety proves they have not been formed by subsidence as some writers of theories of the earth have pretended, for then they should always have been in circular strata, and concentric. It proves too that they have not been formed by the rotation of the earth on it's axis, as might have been suspected had all these strata been parallel with that axis. They may indeed have been thrown up by explosions, as Whitehurst supposes, or have been the effect of convulsions. But there can be no proof of the explosion, nor is it probable that convulsions have deformed every spot of the earth. It is now generally agreed that rock grows, and it seems that it grows in layers in every direction, as the branches of trees grow in all directions. Why seek further the solution of this phaenomenon?"[15]

By the end of the century, two distinct schools of thought had developed concerning the forces creating and shaping the earth. One of the key technical issues concerned the formation of granites. Either the agency had been heat (the Vulcanist theory) or water (the Neptunist theory). Divisions over these theories were to excite American and European geologists for decades to come.[16] Jefferson kept abreast of all these geological developments, but his interest in this subject declined after he completed *Notes* and when he lived in Paris. He failed to find, either from his own thinking or from the researches of others, any solution to these problems, and the conservatism required by his adherence to the biblical account of creation prevented him from accepting more modern scientific views. As time went by, he adopted a rather peevish and even anti-intellectual approach to it all. He continued to argue that mineralogy and geology were central to a young person's education, but he rejected theory. In 1805 he wrote to the French immigrant anthropologist and geographer C. F. C. de Volney that he no longer "indulged" in "geological inquiries": "the skin-deep scratches which we can make or find on the surface of the earth, do not repay our time with as certain and useful deductions as our pursuits in some other branches."[17]

Jefferson made an even more negative statement about geological theory in 1826, in a letter advising Dr. John Emmett on a curriculum for the new University of Virginia: "The dreams about the modes of creation, inquiries whether our globe has been formed by the agency of fire or water, how many millions of years it has cost Vulcan or Neptune to produce what the fiat of the Creator would effect by a single act of will, is too idle to be worth a single hour of any man's life."[18] The reference might be to the work of James Hutton in Scotland, who proposed in 1785 a more detailed version of Hooke's idea of cycling processes in the earth's crust.[19] Hutton had set out to discover, from detailed observations of geological formations, the age of the earth. For example, one could measure the rate at which new sediments (from erosion) were laid down and measure the thicknesses of old sedimentary rocks. He failed in that task but laid down the foundation of modern geology.

Parenthetically, however, it is worth noting that Benjamin Franklin had already had a remarkable insight. It throws into sharp focus the difference between the way Jefferson's mind, always reasoning from facts, worked and the way Franklin's did. Franklin's insight was pure speculation. He imagined that "the internal part [of the earth] might be a fluid" and that the solid crust "might swim in or upon that fluid. Thus the surface of the earth would be a shell, capable of being broken and disordered by any evident movements of the fluid on which it rested in which the fundamental movements of the earth are driven by its inner heat."[20] This idea is remarkably like the modern theory of plate tectonics, according to which convection currents in the earth's semi-liquid mantle cause sections of the outer crust (plates) to move around the surface, bumping into each other and being variously forced upward, sideways, or downward. No doubt this interpretation will eventually be superseded by the work of future geologists.

CHAPTER EIGHT

On Fossils and Extinction

WHEN *NOTES ON THE STATE of Virginia* was written, Jefferson had not actually seen a live elephant. The first one, an Asian elephant, was brought to America in 1796, and the next year Jefferson paid five cents to see it on display on Market Street, Philadelphia. Nonetheless, in a genuinely innovative contribution to American paleontology, he had concluded that the Siberian mammoth and the American mastodon were species of the elephant family adapted for cold climates and that the African and the Asian elephants were quite different species adapted for heat. In an early exercise in biogeography, he even defined the ranges of the two kinds as being split at around 36½ degrees latitude north.

But Jefferson was involved in politics and diplomacy for almost a decade after completing *Notes* and had little time for studying fossils until he resigned as Washington's secretary of state in 1793. Then, in April 1796, he received a tantalizing letter from his friend John Stuart in Greenbrier County (then part of Virginia, now West Virginia) about another enormous American mammal. The letter began: "SIR, Being informed you have retired from public Business and returned to your former Residence in Albemarle, and observing by your Notes your very curious desire for Examining into the antiquitys of our Country, I thought the Bones of a Tremendious animal of the Clawed kind lately found in a Cave by some Saltpetre manufacturers about five miles from my House might afford you some amusement, have therefore procured you such as were saved . . . I donot remember to have seen any account in the History of our Country, or any other of such an animal which probabelly was of the Lion kind."[1]

When the bones arrived they were indeed intriguing, consisting of parts of an enormous limb, the foot of which had claws. The keratinous parts were missing, but still the claws were huge. Stuart reported having seen one that was eight inches long and promised to obtain for Jefferson a massive thigh bone that had been seen in the cave. Jefferson eagerly launched into the project of describing the remains and preparing a paper on them for the American Philosophical Society, to whom he planned to donate the remains. He proceeded at first on the assumption that this *Megalonyx*, or "great-claw," was indeed a kind of enormous American lion, as Stuart had suggested. Its claws were at least three times longer than the African lion's. His old friend Daubenton in Paris had recently published an account of an African lion skeleton, complete with detailed measurements, so Jefferson felt quite sure of his ground, but he realized that measurements of the claws alone would not make the case for the superior size of *Megalonyx*. So he wrote to Stuart asking for news about the promised thigh-bone.

"My anxiety to obtain a thigh bone is such that I defer communicating what we have to the Philosophical society in the hope of adding that bone to the collection. We should then be able to fix the stature of the animal without going into conjecture and calculation as we should possess a whole limb from the haunch bone to the claw inclusive. Whenever you announce to me that the recovery of a thigh bone is desperate, I shall make the communication to the Philosophical society."[2] (While he was, in the modern sense, "desperate" to get this thigh bone, Jefferson is here using the word in its contemporary sense of "without hope.") Jefferson did not want to complete his manuscript without examining the crucial thigh bone because it would nail down (so to speak) the question of size. In Jefferson's disputes with the Comte de Buffon, size was all-important; this new animal had to be confirmed as another huge, preferably ferocious emblem of American might and strength. In fact, modern estimates put the weight of an adult *Megalonyx* at around eight hundred pounds.

There is no doubt that Jefferson was led astray by Stuart when the latter suggested in his letter that the animal was "probably of the Lion kind." He almost allowed himself to be carried away by his enthusiasm for this "lion." However, after he finished writing his paper, he came across a drawing of a new animal, *Megatherium*, discovered by Europeans in South America

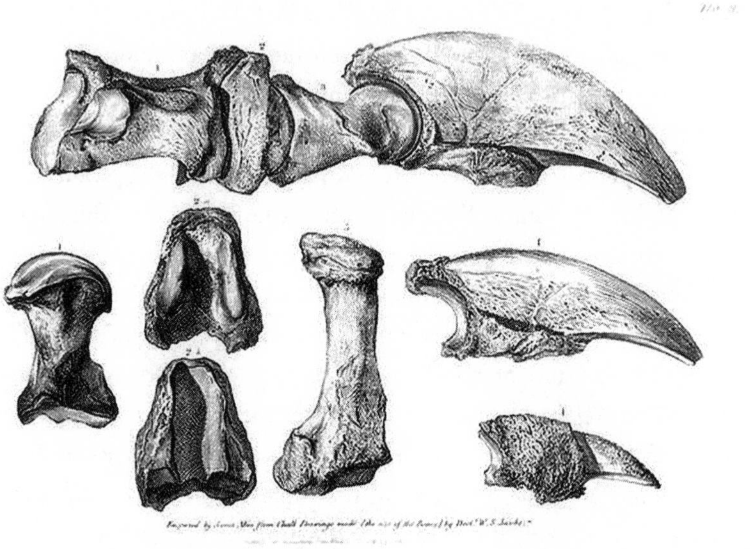

This drawing of the *Megalonyx* remains from (West) Virginia was made by Caspar Wistar for his monograph in the *Transactions of the American Philosophical Society* (1799), which accompanied Jefferson's less formal description of the remains. Courtesy of the American Philosophical Society.

and instantly recognized what his great-claw was.[3] Caspar Wistar, the Philadelphia anatomist, who had begun a more technical description—actually it was a quite brilliant piece of forensics showing among other things that the claws were not retractable—agreed.[4] The great-claw was not a lion but a ground sloth, not symbolic of strength and vigor so much as of, well, sloth.[5]

Jefferson quickly revised his manuscript, adding a page-long postscript about *Megatherium.* It was published in 1799, along with Wistar's account, in the fourth volume of the society's *Transactions.*[6] Even a casual reading shows how reluctant he must have been to abandon his lion idea. In fact, Jefferson still seemed to cling to the idea that the skeleton would be eventually revealed as that of a giant lion after all.

With his discussion of the mastodon in *Notes,* Jefferson had been the first to write analytically about American fossils. With his account of the great-claw, he became the first person to describe and name an American fossil mammal. He promoted a popular interest in American fossils by his encouragement of Charles Willson Peale's excavations in New York State. With his emphasis on the unique size of both the mastodon and the great-claw, he stands at the beginning of a line that ends with the discovery of the tremendous dinosaurs and giant mammals of the American West. We can only guess how fascinated he would have been to see these marvels. If he had lived to see dinosaurs, he would have felt even more vindicated in his view that American wildlife was superior to anything in the Old World. All these accomplishments would seem to qualify Jefferson as the founder of American vertebrate paleontology, and many people have claimed this distinction for him. But here, as in so many other aspects of his extraordinary life, controversy has arisen.

There are two immediate problems. Jefferson's use of the scientific method can be criticized because instead of gathering facts impartially and then analyzing them (the way his idol Bacon urged), Jefferson had worked a priori. He had started with his lion idea and then tried to prove it. Worse, when that identification was disproven, he still did not completely relinquish the theory. The second criticism, more serious to modern commentators, is that he did not believe in extinction, and that view seems to have been based in large part on religious rather than scientific grounds.[7]

Both charges are true. Even in the letter that Jefferson wrote to Stuart asking about the thigh bone, he took time (almost defensively) to write out an argument against extinction: "I cannot however help believing that this animal as well as the Mammoth are still existing. The annihilation of any species of existence is so unexampled in any parts of the economy of nature which we see, that we have a right to conclude, as to the parts we do not see, that the probabilities against such annihilation are stronger than those for it."[8]

The simplest statement of fact, he wrote, was that "the bones exist; therefore the animal has existed . . . If this animal then has once existed, it is probable . . . that he still exists." In *Notes* he had made a similar argument against the extinction of the mastodon. "Such is the oeconomy of nature, that no instance can be produced of her having permitted any one race of her animals to become extinct; of her having formed any link in her great work so weak as to be broken."[9] (This was a reference to the classical concept of a Great Chain of Being.)[10]

The response to Jefferson's modern critics, however, is that *for his time,* Jefferson was not being unscientific with respect to extinction. His argument against extinction was both straightforward and complex, not mere wishful thinking. That the mastodon and the great-claw were still extant could be seen as a formal scientific hypothesis, and not only was this hypothesis testable but, like any scientist (or lawyer), Jefferson argued from evidence. Jefferson devoted four pages of his great-claw paper to the extensive evidence in Indian legends and in travelers' reports of the existence of unknown, fearsome predators in the West. There were (and he really wanted to think there were) valid reasons to suspect that the great-claw was alive.

Reports about monstrous animals roaming the American West continued to surface after Jefferson had finished his *Megalonyx* paper. A friend wrote to Jefferson in September 1798, "Some circumstances have lately been related to me which in my opinion go far in support of your conjecture that the species of *Animal whose* Bones *were found in Green briar,* still exists in the Western Country . . . far up the Misouri river . . . [An] Animal is found of a brown colour, much larger than a Bear, of astonishing strength, activity, & fierceness . . . a Nail [was] taken from the Claw of one killed by the party of Indians to which he belonged but not before it had torn several of them into pieces. This Nail or horny part of the Claw is said to measure six inches in length."[11] Such reports continued well into the next century. For instance a very similar account was contained in a letter of 1836 from the naturalist John Kirk Townsend to Samuel G. Morton.[12]

Jefferson never changed his mind about the extinction of species, although he understood that races and populations (like the wolf in Britain or American Indian tribes and languages) might become extinct. As he wrote to John Adams in 1823, "We see . . . evident proofs of the necessity of a superintending power to maintain the Universe in it's course and order . . . Certain races of animals are become extinct; and, were there no restoring power, all existence might extinguish successively, one by one, until all should be reduced to a shapeless chaos."[13]

It is not clear just what Jefferson meant by "restoring power." Earlier in the same letter, he had written that he believed that the world had been created by an ultimate cause (he is careful not to say God) who was "a fabricator of all things from matter and motion, their preserver and regulator while permitted to exist in their present forms, and their regenerator into *new and other forms*" (emphasis added).[14] By this date, the early evolutionary ideas of Erasmus Darwin (grandfather of Charles Darwin) had long since entered the literature.[15] Jefferson had had copies of his books in his great library. It is possible that Jefferson was leaving a door open for evolution ("transmutation of species") when he wrote of "new and other" forms, but there are no other such references. It is more likely that he was referring, in this letter to Adams, to the power of nature to restore the loss of races and varieties within species.

Meanwhile, almost all other scholars had accepted that extinction was a fact and that the mastodon, at least, was gone from the earth. Benjamin Smith Barton wrote Jefferson a long letter on the subject in 1810, the occasion being his receipt of new information and detailed drawings of the Siberian mammoth. "Of all the subjects of animal natural history, there is not one more interesting than that which relates to the character and history of those vast organized bodies,—many of them, too, endowed with an immense portion of intelligence,—which the God of Nature had created; and after suffering them to grow and exist through the ages, unknown ages of time, has, at length, entirely removed from the earth; not merely as individuals, but as *species*. There is something awful about the consideration of this subject: and yet thee very subject is admittedly calculated to display to us the wisdom, as well as power, of him who formed all things. The harmony of nature is not, in the smallest degree, disturbed by the total destruction of what may have deemed *necessary* integral parts of a common whole. Nor is this business of the extinction of species at an end. That which has already taken place, with respect to species of Elephant, Rhinoceros, and other vast families of animals, will unquestionably take place with respect to many of the families of animals—and none, I think, more rapidly than in that portion of it which we inhabit."[16]

Robert Livingston of New York had written to Jefferson nine years earlier suggesting an ecological explanation of the extinction of mastodons. Mastodons had served, he thought, to keep down other "gramivorous" mammals like deer and bison, but "when a race of Savage men were transplanted into our forests they were no longer necessary." He went on to state that extinction was then inevitable for the Indians also, putting it bluntly: "And do we not see, Sir, in the gradual but certain annihilation of these very red Children something like a similar disposition of Providence?"[17]

On the face of it, it might seem odd to the point of absurdity that Jefferson denied extinction when he had the fossils in front of him. However, a careful rereading of Jefferson's writings on what we call fossils shows that for two hundred years historians and biographers, virtually without exception, have missed an important point. Despite criticism of his methods, they have referred admiringly to Jefferson's love of fossils, to his great-claw paper, to the collection of mastodon fossils that he kept at Monticello, and to his encouragement of an interest in paleontology by others. But in fact *Jefferson did not refer to the mastodon and the great-claw remains as fossils as we now understand that term.*

Several of Jefferson's correspondents used the term "fossil" (or "extraneous fossil") in its present sense of organic remains buried within the rock and mineralized (petrified). But the word "fossil" does not appear anywhere in Jefferson's great-claw paper. He never used the word "fossil" to refer to great-claw (*Megalonyx*) or mastodon remains. He did not use it anywhere in *Notes,* nor does it occur in any of his published letters. For Jefferson, the remains of the great-claw and the mastodon were always "bones" or "big bones." The ancient remains of shells were always "petrified shells." We are then left to ask, Why did Jefferson not employ the word "fossil," and what did he think constituted a fossil? This semantic puzzle offers a tantalizing example of how difficult it is to be sure of a reading of Jefferson's works and letters because of differences in word usage then and now. Indeed, usages were constantly changing in his lifetime, just as they are today.

The word "fossil," in Jefferson's time, was undergoing a shift in meaning. As a noun or an adjective, it had originally meant anything that was "dug up" (the word "fossil" comes from the Latin for "to dig"). In Europe, by the mid-eighteenth century a distinction was made between two kinds of objects found in the ground. Inorganic objects taken from the ground were usually called "natural fossils." They included salts, bitumen, coal, minerals, gemstones, and metallic ores. The more restricted term "extraneous fossil" (or, less frequently, "adventitious fossil") was reserved for organic remains, the preserved remnants of once living animals that were petrified (a term now considered synonymous with "fossilized").[18] Extraneous fossils were implicitly old (although the question of age was limited where the belief persisted that the age of the world was only a few thousand years).[19] By the end of the eighteenth century the term "fossil" was almost exclusively used in this latter sense; eventually the word "extraneous" was dropped, and everything once termed a "natural fossil" acquired its own separate description as a mineral, salt, ore, and so on.

Jefferson was familiar with this newer set of definitions. In the same issue of the American Philosophical Society *Transactions* where his great-claw paper appeared in 1799, there was an essay on mastodon and other remains from Big Bone Lick, calling them "extraneous fossils." Back in 1768, Fothergill had described a "fossil Allegator" from eastern England that was clearly both transformed into rock and entombed within the rock.[20] When the same author described, in 1767, the mastodon teeth that George Croghan had sent to London, the title of his paper for the Royal Society included the words "some very large Fossil Teeth." The following month, William Hunter wrote about the same sets of remains, using the words "fossil ivory" and "fossil teeth." In 1784, Big Bone Lick fossils collected by a Major Craig were displayed at the American Philosophical Society. In the minutes of the meeting they were described as "petrified bones," and a parallel article in *Columbia Magazine* referred to the exhibition of "fossil tusks."[21]

Americans were somewhat slower than Europeans, however, in adopting the more refined usages. When the American Philosophical Society, in 1771, proposed to create a "cabinet," or museum, they referred to collections of "all Specimens of Natural Productions, whether of the ANIMAL, VEGETABLE or FOSSIL Kingdoms."[22] A decade later, Pierre Eugene du

Simitiere advertised his "American Museum" in Philadelphia as containing both "Natural Curiosities" and "Artificial curiosities." The latter ranged from portraits to clever machines. The former included "Marine Productions" (fishes and the like), "Land Productions" (birds and insects, for example), "Fossils" (used in the old sense of minerals and salts), and "Petrifications." This last category included both what we now call fossils (such as bones, shells, teeth, corals) and a number of inorganic items, such as "fossil substances produced by the eruptions of Volcanos." As late as 1792, when Charles Willson Peale proposed *his* new museum for Philadelphia, he wrote that it would include "the fossil kingdom, comprehending the earths, minerals, and other fossil matters, which include petrefections."[23] Even in reporting on his excavations of mastodons in New York State, Peale rarely used the word "fossil" except in a general sense. He told Jefferson that he had sent his son Raphaelle to the "western country" "to get all the fossil Bones he can."[24] But Peale never used the word when referring to a particular bone.

A rigorous application of these older ways of classifying nature means that a particular item had to be either a bone or a "petrefection," but it could not logically be both.

The first questions to be asked about Jefferson's failure to use the word "fossil" are whether he thought the mastodon and *Megalonyx* bones had been "dug up" and whether they were petrified. It is not clear that Jefferson thought the great-claw remains were fossils in the technical sense of having been "dug up." They appeared to come from the cave floor. The mastodon remains at Big Bone Lick came from deposits of a peaty marsh rather than rock. The first specimens had been found on the surface, although, when Meriwether Lewis visited Big Bone Lick in 1803, he was explicit that a collector had now to dig down ten feet or more to find any. Previous collectors, like Dr. John Gosforth, had picked the place clean of any remaining bones, tusk, and teeth that had once been exposed by trampling bison or erosion. (By that time also, Charles Willson Peale had excavated mastodon skeletons from the flooded "morasses" of marl in New York State.)

Not only had the great-claw remains not been fully encased within rock but Jefferson was explicit that they were not fully petrified.[25] In his formal account of *Megalonyx* he explained their condition as the result of lying in the saltpeter cave. "The nitrous impregnation of the earth together with a small degree of petrification had probably been the means of their preservation."[26] When Meriwether Lewis stopped at Big Bone Lick to look for incognitum remains in 1803 on the way to lead the expedition of exploration with William Clark, he wrote to Jefferson, "Not any of the bones or tusks which I saw were petrifyed, either preserving their primitive states of bone or ivory; or when decayed, the former desolving into earth intermixed with scales of the header or more indissoluble parts of the bone, while the latter assumed the appearance of pure white chalk."[27] Peale thought his New York materials were even less petrified than the Big Bone Lick materials, although there were many highly petrified shells, nuts, and corals in adjacent beds.[28]

One might think that the great-claw and mastodon remains would at least be denser and heavier than comparable recent bones and teeth. Indeed, mastodon molar teeth are heavy. However, even that turns out to be a nuanced issue. Comparison in the laboratory of some of Jefferson's mastodon teeth from Big Bone Lick and Asian elephant teeth collected in the nineteenth century (and therefore very dry) of the same size reveals, somewhat surprisingly, that they had approximately the same relative density: approximately 2.0 grams per cubic centimeter. This is not to say that Jefferson had access to recent elephant teeth for comparison. It merely indicates that answers to questions about what Jefferson knew as fact are not simple.

The one exception to Jefferson's otherwise blanket avoidance of the word "fossil" perhaps "proves the rule." In 1807, Jefferson commissioned William Clark to make a large collection of bones from Big Bone Lick. When formally commissioning Clark in 1807, Jefferson had referred only to "bones." When he thanked him for the results two years later, he also wrote "bones."[29] In his Memorandum Book entry for February 9, 1808, however, he wrote, "Gave ord. on bank US. In favr. Genl. Wm. Clarke for expenses digging fossil bones."[30] There is a similar notation for November 2 of that year where he used the spelling "fossile."[31] At the very least, in these entries Jefferson was admitting that the bones had been "dug up."

Just as a great deal about Jefferson's life and personal philosophy is paradoxical and arouses great passions, his devotion to science and its methods is here both reinforced and questioned. The simplest explanation for Jefferson's failure to use the word "fossil" when writing formally may be that he had not yet adopted the new terminology. But he may also not have wanted to commit himself as to his bones' technical status. He may have been unsure whether they qualified as fossils under the new contemporary definitions of the term.

Cutting across any scientific considerations is a religious one. Jefferson did not believe that God would have allowed any of the creatures he had created to become extinct. To believe that would be like admitting that God had made a mistake or, at least, had made "two jobs" of something that the Bible said was a single six-day event. Jefferson's avoidance of the word "fossil" helped reinforced his stance on extinction. Since Jefferson did not even use the word "fossil" in connection with the petrified shells found in the Blue Ridge Mountains, it seems likely that Jefferson associated the term with extinction. The strongest statement he made on the subject was in *Notes:* "Near the eastern foot of the North mountain are immense bodies of *Schist,* containing impressions of shells in a variety of forms. I have received petrified shells of very different kinds from the first sources of the Kentucky, which bear no resemblance to any I have ever seen on the tide-waters."

If, on the one hand, the mastodon and the *Megalonyx* remains were petrified fossils in the modern sense, that would have increased many times the probability that the animals were extinct and had been dead for a very long time. On the other hand, if the bones were *not* technically fossils in the sense of either dug-up or petrified, they were much less likely to be from an extinct animal or even to be very old. In this connection it is useful to recall that when Jefferson discussed various theories to account for the presence of a supposedly warm-weather elephant in the "frozen zone," he rejected the idea that the obliquity of the earth's axis had pronouncedly changed, altering the climate from warm to cold, because that would require the (unacceptable) elapse of a quarter of million years or more.

It may be that Jefferson's fascination with fossils, so well known in his time and ours, created yet another set of conflicts that he could not resolve. By using the term "bones," Jefferson was able to proceed noncommittally with respect to extinction. It could have been a deliberate compromise. That is, he might have realized (or suspected) that the remains were fossils in every modern sense of that term but did not use the word because doing so would allow others to see him as endorsing extinction. Such a rhetorical strategy would have been devious but not dishonest. After all, he thought he had good anecdotal evidence for the existence of living *Megalonyx* and mastodons.

There remains a simple irony: Jefferson helped found the science of American paleontology while rejecting both of what are now its basic premises—fossils and extinction.

PART THREE

They, the People

CHAPTER NINE

Europe and the Peoples of America

"LET US NOW EXAMINE WHY the reptiles and insects are so large, the quadrupeds so small, and the men so cold, in the New World." These few words, written in 1761, capture the arrogance of Buffon's attitude toward the Americas. Of all Buffon's writings on nature, none seems to have been more provocative or more biased by philosophical and hypothetical considerations than his view of the native peoples of the Americas.

Jefferson's anger at Buffon's dismissal of Native Americans is shown by the fact that he copied Buffon's remarks into *Notes on the State of Virginia* for his readers to see. "Although the savage of the new world is about the same height as man in our world, this does not suffice for him to make an exception to the general fact of the reduction of living nature in all that continent. The savage is feeble, and has small organs of generation; he has neither hair nor beard, and no ardour for his female; although more agile than the European because he has the habit of running, he is, however, much less strong in body; he is much less sensitive, and yet more timid and cowardly; he has no vivacity, no activity of mind . . . Nature, by refusing him the power of love, has treated him worse and lowered him deeper than any animal."[1] The original passage in *Histoire Naturelle* continued in this vein.

For Buffon, the animals and native peoples of the Americas were one and the same in their undeveloped, primitive, or degenerate condition, and he consigned them to a lesser status in the natural scheme of things. In answering him, as least as far as the indigenous people of Virginia were concerned, Jefferson felt on sure ground. He had grown up knowing the Indians who traveled across his father's lands on their way to treaty with the government in Williamsburg. He later recalled in almost romantic terms that "I knew much the Great Outassete, the warrior and orator of the Cherokees . . . I was in his camp when he made his great farewell oration to his people, the evening before his departure for England. The moon was in full splendour, and to her he seemed to address himself in his prayers for his own safety on the voyage, and that of his people during his absence. His sounding voice . . . and the solemn silence of his people at several fires, filled me with awe and veneration, altho' I did not understand a word he uttered."[2]

In *Notes on the State of Virginia,* and especially in the added appendix by Charles Thomson, Jefferson defended the American Indian vehemently. "He is neither more defective in ardor, nor more impotent with his female, than the white reduced to the same diet and exercise: that he is brave, when an enterprize depends on bravery; education with him making the point of honor consist in the destruction of an enemy by stratagem, and in the preservation of his own person free from injury; or perhaps this is nature; while it is education which teaches us to honor force more than finesse: that he will defend himself against an host of enemies, always chusing to be killed, rather than to surrender, though it be to the whites, who he knows will treat him well: that in other situations also he meets death with more deliberation, and endures tortures with a firmness unknown almost to religious enthusiasm with us: that he is affectionate to his children, careful of them, and indulgent in the extreme: that his affections comprehend his other connections, weakening, as with us, from circle to circle, as they recede from the center: that his friendships are strong and faithful to the uttermost extremity: that his sensibility is keen, even the warriors weeping most bitterly on the loss of their children, though in general they endeavour to appear superior to human events: that his vivacity and activity of mind is equal to ours in the same situation; hence his eagerness for hunting, and for games of chance. The women are submitted to unjust drudgery. This I believe is the case with every barbarous people. With such, force is law. The stronger sex therefore imposes on the weaker."[3]

Benjamin Rush made a note in his Commonplace Book that is relevant here, even if it was unfashionable at the time: "The Indian Savages oblige their women only to work. Among civilized nations, the women oblige the men only to work. The men among the former, and the women among the latter consider the opposite sex made only to administer to their comfort without any cooperation on their part. Both are wrong. Men and women were made to work together in different ways."[4]

Jefferson admitted that Indians "raise fewer children than we do. [But] the causes of this are to be found, not in a difference of nature, but of circumstance. The women very frequently attending the men in their parties of war and of hunting, child-bearing becomes extremely inconvenient to them. It is said, therefore, that they have learnt the practice of procuring abortion by the use of some vegetable; and that it even extends to prevent conception for a considerable time after." Jefferson added to his facts the observation that the "same Indian women, when married to white traders, who feed them and their children plentifully and regularly, who exempt them from excessive drudgery, who keep them stationary and unexposed to accident, produce and raise as many children as the white women."

As for the Indians being beardless, there was abundant evidence that the Indians had plenty of hair. In 1724, Hugh Jones, who despised the Indians, wrote in his history of Virginia that "their hair is very black, coarse and long."[5] Mark Catesby, in his travels in the Carolinas, had solved the problem of their apparent beardlessness, noting that the "Indians of Carolina are generally tall, and well shap'd . . . their Hair is black, lank, and very coarse . . . The beards are naturally very thin of hair, which they are continually plucking away by the Roots."[6] Jefferson repeated this: "It has been said, that Indians have less hair than the whites, except on the head. But this is a fact of which fair proof can scarcely be had. With them it is disgraceful to be hairy on the body. They say it likens them to hogs. They therefore pluck the hair as fast as it appears . . . Nor, if the fact be true, is the consequence necessary which has been drawn from it. Negroes have notoriously less hair than the whites; yet they are more ardent."

One prime piece of evidence quoted by Jefferson came to acquire considerable fame and some notoriety. This was the "speech of Chief Logan." Logan was the European name of a Cayugan Indian of the Iroquois Confederacy whose family was murdered in 1774 by a party of whites under

"Colonel Cresap" after the series of raids and reprisals that made up "Lord Dunmore's War."[7] Jefferson wrote of Logan's eloquent, even noble, but unrepentant speech that "I may challenge the whole orations of Demosthenes and Cicero, and of any more eminent orator, if Europe has furnished more eminent, to produce a single passage, superior to the speech of Logan, a Mingo chief, to Lord Dunmore, when governor of this state."

Chief Logan is reported to have said: "I appeal to any white man to say, if ever he entered Logan's cabin hungry, and he gave him not meat; if ever he came cold and naked, and he clothed him not. During the course of the last long and bloody war, Logan remained idle in his cabin, an advocate for peace. Such was my love for the whites, that my countrymen pointed as they passed, and said, 'Logan is the friend of white men.' I had even thought to have lived with you, but for the injuries of one man. Col. Cresap, the last spring, in cold blood, and unprovoked, murdered all the relations of Logan, not sparing even my women and children. There runs not a drop of my blood in the veins of any living creature. This called on me for revenge. I have sought it: I have killed many: I have fully glutted my vengeance. For my country, I rejoice at the beams of peace. But do not harbour a thought that mine is the joy of fear. Logan never felt fear. He will not turn on his heel to save his life. Who is there to mourn for Logan?—Not one."

His speech was almost an elegy for the passing of an era in Indian life. It was often quoted before Jefferson included it in *Notes,* whereupon it became even more famous in both the United States and Europe. Jefferson's uncompromising espousal of Logan's cause when many details were cloudy—even Logan's identity—also became a bone of contention between Jefferson and Cresap's family.

Notes presents a romantic picture. Jefferson did not always view American Indians through a rosy lens, however, or cast upon them an entirely benevolent eye. He could be cold and dispassionate, almost in the same breath as he was admiring. In the letter to John Adams in which he rhapsodized about Chief Outassete, he wrote that the Cherokees in Virginia now numbered "abut 2000" and were "Far advanced in civilisation. They have good Cabins, inclosed fields, large herds of cattle and hogs, spin and weave their own clothes of cotton, have smiths and other of the most necessary tradesmen . . . Some other tribes were advancing in the same line." But, he warned ominously, while those Cherokees who became farmers and acculturated would prosper, the "backward will yield, and be thrown further back. These will relapse into barbarism and misery, lose numbers by war and want, and we shall be obliged to drive them, with the beasts of the forest into the Stoney mountains."[8]

Once again we see two sides of Jefferson: the romantic, almost dreamily remembering the Indians of his youth, and the hard-nosed legislator. It is as if he could distinguish between Indians as individuals and groups and Indians as inconvenient obstacles to western expansion and settlement. Anthony Wallace in his now-classic book *Jefferson and the Indians* describes in painful detail how Jefferson's principles and policies played a significant role in preparing public opinion for and precipitating the Indian Removal Act of 1830, passed during the presidency of Andrew Jackson, after which the "Five Civilized Tribes" were forcibly relocated from the American South, taking the "Trail of Tears" (the Cherokees ended up in present-day Oklahoma).[9] There was considerable pressure for removal of Indians from the South during Jefferson's presidency, and as Wallace shows, Jefferson urged their removal. After 1809, when Jefferson left office, the pressure lessened before it built up again. With Jefferson, the irony seems inevitable. Part of his reason for acquiescing to Indian removal was support for the rights of states to act individually.[10] The eventual forced relocation was by federal act.

As a scholar, Jefferson had a strong scientific interest in the question of the origins of the various American Indian groups. He thought that some answers might be found in a study of their languages; and looking for terms held in common would lead to the discovery of a sort of genealogy of tribes. He thought that the differences in language, which were often extreme, even between closely neighboring tribes, might have been reinforced by their need for cultural separation, one from another. (There is a strong hint here of the theories of Charles Darwin a century later.)

"We are told that the Powhatans, Mannahoacs, and Monacans, spoke languages so radically different, that interpreters were necessary when they transacted business. Hence we may conjecture, that this was not the case between all the tribes, and probably that each spoke the language of the nation to which it was attached; which we know to have been the case in many particular instances. Very possibly there may have been antiently three different stocks, each of which multiplying in a long course of time, had separated into so many little societies. This practice results from the circumstance of their having never submitted themselves to any laws, any coercive power, any shadow of government . . . It will be said, that great societies cannot exist without government. The Savages therefore break them into small ones.

"It is to be lamented then, very much to be lamented, that we have suffered so many of the Indian tribes already to extinguish, without our having previously collected and deposited in the records of literature, the general rudiments at least of the languages they spoke. Were vocabularies formed of all the languages spoken in North and South America, preserving their appellations of the most common objects in nature, of those which must be present to every nation barbarous or civilised, with the inflections of their nouns and verbs, their principles of regimen and concord, and these deposited in all the public libraries, it would furnish opportunities to those skilled in the languages of the old world to compare them with these, now, or at any future time, and hence to construct the best evidence of the derivation of this part of the human race."

Notes also contains important contributions that Jefferson made to American Indian archaeology and anthropology. Barbé-Marbois had originally asked for information about "Indian monuments." In chapter 11 of *Notes,* evidently repeating a passage from his original reply to the Frenchman, Jefferson wrote: "I know of no such thing existing as an Indian monument: for would not honour with that name arrow points, stone hatchets, stone pipes, and half-shapen images . . . unless indeed it be the Barrows, of which many are to be found all over this country. These are of different sizes, some of them constructed of earth, and some of loose stones . . . There being one of these in my neighbourhood, I wished to satisfy myself whether any, and which of these opinions were just. For this purpose determined to open and examine it thoroughly."

After casual digging produced a series of bones, "I proceeded then to make a perpendicular cut through the body of the barrow, that I might examine its internal structure. This passed about three feet from its center, was opened to the former surface of the earth, and was wide enough for a man to walk through and examine its sides. At the bottom, that is, on the level of the circumjacent plain, I found bones; above these a few stones, brought from a cliff a quarter of a mile off, and from the river one-eighth of a mile off; then a large interval of earth, then a stratum of bones, and so on."

By excavating this Indian mound, Jefferson became the first to undertake systematic archaeological excavation, strata by strata. His conclusion that the site was for burial was confirmed by the fact that "about thirty years ago," a party of Indians coming through the region "went through the woods directly to it, without any instructions or enquiry, and having staid about it some time, with expressions which were construed to be those of sorrow."

The fact that Buffon had insisted that all the native peoples of the Americas were weak, cowardly, and uncivilized raised a serious question about the sources upon which Buffon was relying. He had completely ignored, for example, the reports of the Spanish explorers to South and Central America of the seventeenth century, who brought back accounts of elaborate cities, complex architecture, and sophisticated gold objects and jewels. He evidently relied more upon the opinions of modern Spanish and Portuguese authors than upon French reports. Robert La Salle and Samuel de Champlain, who had explored in North America, had told a different story about the culture of North American Indians, and Buffon seems also to have ignored the eager French recruitment of native North Americans to fight on France's side in the French and Indian War Although the Indians might have switched sides when expedient, no one doubted their skill as fighters.

The sources of Buffon's negativity about the Americas are to be found in both the scientific and social-political spheres.[11] Many contemporary European authors had a distinct animus against the Americas, both the decadent regimes of South America and the upstart colonies of British

North America (soon to be United States and Canada). Buffon's own views were not purely a matter of scientific fact. When it came to the status of "American" Indians he had strong philosophical opinions. For him, the Indians of North and South America were not the noble savages envisaged by the admirers of Jean-Jacques Rousseau. They were either of Eurasian stock and had degenerated because of the awful climate, or they were genuinely in a primitive brutish state. For Buffon they were classically "they"—"les petites nations sauvage de l'amérique"—and distinctly not "us": "nos grand peuples civilizes."[12]

Obviously Buffon's writings were not directed against the United States per se. North America east of the Mississippi was still British when Buffon wrote in 1761. But a strong case can be made for European resentment of the fact that, having taken control of French Canada after the French and Indian War, British North America now had limitless lands, valuable natural resources, and strong, confident people. The "new" continent, moreover, was daily attracting more and more European immigrants. North America had grown stronger and stronger during the middle of the eighteenth century, and Europe weaker.[13] As Hannah Adams wrote in her history of New England (1799), "At this period the arms of Great Britain had recently been successful in every part of the globe. Power, however, like all things human, had its limits; and there is an elevated point of grandeur which seems to indicate a descent. The kingdoms of Europe looked with a jealous eye upon Britain, after the acquisition of such immense power and territory . . . while the ideas of liberty and independence . . . were increased."[14] There was good reason why minor aristocrats, which all the leading French scientists were, would want to denigrate America and its educated, free populace.

Although Buffon did not credit any of his sources in the 1761 volume of *Histoire Naturelle,* where he first denigrated the Americas, it was clear to Jefferson that Buffon had principally taken his views from the writings of Don Antonio d'Ulloa (1716–1795), a Spanish general and colonial administrator who traveled in Ecuador, Peru, and Louisiana. Two other prominent Europeans, Corneille de Pauw (1739–1799) and Abbé Guilaume Thomas François Raynal (1713–1796), contemporaries of Buffon's, seem also to have influenced him. Both were frankly politically biased, writing polemics against colonialism, slavery, and exploitation. Their focus was the iniquity of European rule in South and Central America—in slavery, colonization, exploitation, and genocide. For them, there was little or nothing good about South America, except what Europeans had stolen, and the state of the wildlife confirmed the odious condition of the country. It was de Pauw whom Buffon quoted in his 1778 volume to the effect that everywhere in the New World, just below the surface, the soil was always frozen. Travelers in the extreme northern and southern regions had observed the frozen soil, but the statement shows de Pauw's and Buffon's willing ignorance—since they had never been to the New World—of everything in between.

"Don Ulloa's testimony is of the most respectable," said Jefferson. "He wrote of what he saw. But he saw the Indian of South America only, and that after he [the Indian] had passed through ten generations of slavery. It is very unfair, from this sample, to judge of the natural genius of this race of men: and after supposing that Don Ulloa had not sufficiently calculated the allowance which should be made for this circumstance, we do him no injury in considering the picture he draws of the present Indians of S. America as no picture of what their ancestors were 300 years ago. It is in N. America we are to seek their original character: and I am safe in affirming that the proofs of genius given by the Indians of N. America, place them on a level with Whites in the same uncultivated state."

Jefferson had total contempt for de Pauw. "Paw, the beginner of this charge, was a compiler from the works of others; and of the most unlucky description; for he seems to have read the writings of travellers only to collect and republish their lies. It is really remarkeable that in three volumes 12mo. of small print it is scarcely possible to find one truth, and yet that the author should be able to produce authority for every fact he states, as he says he can."[15]

While Central and South America presented a threat of colonial decadence, in North America Yankee independence presented a direct (if still distant) challenge to European power. De Pauw was no doubt right to condemn southern colonial regimes, but as the French explorer and writer Antoine-Joseph Pernety, who challenged his ideas on American degeneracy, pointed out, de Pauw and Buffon were wrong to state that there were few people in South America before 1492 and that they had no culture.

Disparaging Native Americans as feeble and unintelligent was as old as the first accounts of Columbian-era explorers, such as Oviedo's *Historia general y natural de las Indias* (1535) and *Sommario de la natural historia de las Indias* (1526), Antonio de Herrera's *Descripcion de las Indias Occidentales* (1601), Bartolomé de las Casas's *Brevísima relación de la destrucción de las Indias* (1552), and Antonio de Solís y Rivadeneyra's *Historia de la conquista de México* (1684).[16] It resulted in large part from the Europeans' attempts to make the Indians into their slaves. De las Casas, who joined Columbus on his second voyage to Hispaniola, bore the typically conflicted attitude of a European. Generally, Europeans found the "Indians" weak and unwilling to work; the Indians made very poor slaves for working the Europeans' plantations. De las Casas wished to protect these defenseless people (whose weakness and unwillingness might seem understandable in the face of the European guns and whips). His solution was to take the pressure off the Indians by bringing in slaves from Africa. Africans proved sturdier and more submissive: "the labour of one negro was computed to be equal to that of four Indians." As a result, the well-meaning de las Casas, in reducing "one race of men to slavery . . . was consulting about the means of restoring liberty to another."[17]

D'Ulloa and Jorge Juan, in the account of their travels (1735 to 1744) in Ecuador, admitted that "it is no easy task to exhibit a true picture of the customs and inclinations of the Indians, and precisely display their genius and real turn of mind." They found them to be "robust, and of a good constitution" with "remarkable longevity," against which they also said, "Nature . . . begins to decay at the age of thirty, whereas the females rather enjoying a more confirmed state of health and vigour" owing to climate and diet. The weakness of the males was due to "early intemperance and voluptuousness." But the males were also lazy, submissive, "slow but very persevering," drunken, and with little imagination. D'Ulloa concluded that they were basically "little above brutes."[18]

Even so, d'Ulloa reported that de las Casas, who "in 1504 . . . began the war against the Indian inhabitants [of Cartagena, Colombia,] . . . met with greater resistance than they expected; those Indians being a martial people, and valour so natural to them, that even the women voluntarily shared in the fatigues and dangers of war."[19] And, like all travelers to South and Central America, d'Ulloa had to admit the contradictory evidence of the architectural monuments of the people, describing the "still superb ruins . . . of the ancient inhabitants of Peru," together with their copper axes, gold utensils, and exquisitely worked emeralds. At the "palace of the yncas at Quito . . . the dignity of the prince will be absolutely conspicuous, in the prodigious magnitude of the materials, and the magnificence of the structure."[20]

A popular device to weaken the evidence of the Indians' architecture and craftwork, with its implications of sophisticated engineering, complex religious beliefs, and well-developed intellect, was to claim that such accounts had been deliberately exaggerated to make the conquerors appear more heroic. Catesby, for example, commented that the reports of authors such as "Herrera, Solis, and other Spanish Authors" were "enough to excit in us a high Opinion of the Knowledge and Politeness of the Mexicans even in the more abstruse Arts of Sculpture and architecture, those darling branches of the Ancients." But, he suspected, these reports had been overstated to "aggrandize their Achievements in conquering so formidable People, who in reality were only a numerous herd of defenceless Indians, and not still continue as perfect Barbarians as any of their Neighbours."[21] D'Ulloa echoed Catesby, referring to the Peruvians as "barbarians" who "live in debasement of human nature; without law or religion; in the most infamous brutality; strangers to moderation; and without the least control or restraint in their excesses."[22]

D'Ulloa, de Pauw, Raynal, and Pernety all wrote a decade before Jefferson began to work on *Notes on the State of Virginia,* so he was reacting against them as well as Buffon. Just as damaging (and at least as threatening) to Jefferson was the fact that the Scottish historian William Robertson, principal of the University of Edinburgh, in his influential *History of America* (1777–1778), had rehearsed Buffon's, Raynal's, and de Pauw's conclusions for an English-speaking audience. The Scot was soon followed by the Englishman Oliver Goldsmith. No doubt both were alarmed at the number of Britons who were cheerfully leaving home to live in supposedly cold, benighted British America, an America that had rejected and taken up arms against the mother country.

Jefferson read Buffon. He owned a copy of Oviedo. He read and quoted from both d'Ulloa and Raynal, although their books do not appear in the list of his library books. He owned Robertson's *History* and copies of Goldsmith's and Pernety's books. In a letter to Benjamin Vaughan, he referred to "the Lies of de Pauw, the dreams of Buffon and Raynal, and the well-rounded periods of their echo Robertson."[23] He thought Robertson's work was particularly dangerous because he appeared to be writing with authority and firsthand acquaintance with the Spanish sources and only weakly disguised his prejudice with sentences like "As all those circumstances that concur in rendering an inquiry into the state of the rude nations in America are intricate and obscure, it is necessary to carry it on with caution." After making such a nod to evenhandedness, he would, without offering counterevidence, plunge headlong into de Pauw–like denigrations. For Robertson the peoples of the Americas were essentially savages, unredeemed by arts, agriculture, or (in most cases) religion. What could commentators do but catalogue all the weaknesses and inefficiencies that they read about and conveniently neglect the rest?

In Robertson's view no one could seriously describe Americans as civilized in the European sense, even among the "tribe of Natchez, and the people of Bogota [who had] advanced beyond the other uncultivated nations of America in their ideas of religion, as well as in their political institutions." Their temples "were constructed with some magnificence," but their religion was "the most refined species of superstition known in America, and, perhaps, one of the most natural as well as most seducing."[24]

Robertson eagerly repeated the notion that "the beardless countenance and smooth skin of the American seems to indicate a defect of vigour, occasioned by some vice in his frame. He is destitute of one sign of manhood and of strength. Even in climates where this passion usually acquires its greatest vigour, the savage of America views his female with disdain, as an animal of a less noble species . . . Nor is this reserve to be ascribed to any opinion they entertain with respect to the merit of female chastity. That is an idea too refined for a savage, and suggested by a delicacy of sentiment and affection to which he is a stranger."[25]

As to the claim that Europeans found so titillating—that American Indians had feeble organs of generation—the only possible source seems to be something that Buffon's main source, d'Ulloa, wrote. He said that Indians of Ecuador had hair "but no beard; and the greatest alteration occurring by their arriving at the years of maturity is only a few struggling hairs on the chin, but so short and thin, as neither to require the assistance of the razor . . . nor have either males or females any indication of the age of puberty."[26] It may have been this statement that Buffon seized upon to exaggerate into a claim of sexual incompetence and coldness.

As far as most American readers were concerned, Jefferson had answered Buffon and the Europeans in a most satisfactory and conclusive manner. In Europe, however, the debate raged on, especially because of the contributions that its main cast of characters, especially the controversial de Pauw, made to the intellectual history of South America. The debate touched importantly on questions like the existence of the "noble savage," and its influence has been traced to the work of Hegel, Darwin, Kant, and many others, with a return to the American scene via Emerson, Thoreau, and Whitman, among others.[27]

There was one final libel against Americans for Jefferson to deal with, this one put forth by Peter Kalm in his *Travels into North America* and taken as fact by Buffon. Not content with pronouncing that the "cattle degenerate by degrees here, and become smaller," in the very next paragraph Kalm wrote that European people migrated to America "grow sooner old," "do not attain to such an age," and are "far less hardy" than their European contemporaries.[28] This idea was also taken up by de Pauw and Raynal, who claimed that European settlers to North America declined in vigor, generation by generation. Jefferson wrote to Chastellux, "As to the degeneracy of the man of Europe transplanted to America, it is no part of Monsr. de Buffon's system. He goes indeed within one step of it, but he stops there. The Abbé Raynal alone has taken that step. Your knowlege of America enables you to judge this question, to say whether the lower class of people in America, are less informed and less susceptible of information than the lower class in Europe: and whether those in America who have received such an education as that country can give, are less improved by it than Europeans of the same degree of education."[29]

The story has grown up that Franklin, at a dinner in Paris that included Raynal, suggested that the Americans in the room stand up. When they did, they towered over their French hosts, including Raynal, who was himself very short. William Carmichael wrote to Jefferson about what must have been the dinner in question; in fact, neither de Pauw nor Raynal had been present. "I think the Company consisted of 14 or 15 persons. At Table some one of the Company asked the Doctor [Benjamin Franklin] what were his sentiments on the remarks made by the Author of Recherches sur L'Amérique. We were five Americans at Table. The Venerable Doctor regarded the Company and then desired the Gentleman who put the question to remark and to judge whether the human race had degenerated by being transplanted to another section of the Globe. In fact there was no one American present who could not have tost out of the Windows any one or perhaps two of the rest of the Company, if this effect depended merely on muscular force. We heard nothing of Mr. P's work and after yours I think we shall hear nothing more of the opinions of Monsr. Buffon or the Abbé Raynal on this subject."[30] (Franklin himself was taller than average: about five feet nine inches tall.)

Jefferson noted that, for good measure, the Abbé Raynal had complained that "America has not yet produced one good poet, an able mathematician, any man of genius in a single art or science." Jefferson retorted, "When we shall have existed as a people as long as the Greeks did before they produced a Homer, the Romans a Virgil, the French a Racine and Voltaire, the English a Shakespeare and Milton, should this reproach be still true, we will enquire from what unfriendly causes it has proceeded." Once again, he answered opinion with data. "In war we have produced a Washington, whose memory will be adored while liberty shall have votaries, whose name will triumph over time, and will in future ages assume its just station among the most celebrated worthies of the world, when that wretched philosophy shall be forgotten which would have arranged him among the degeneracies of nature. In physics we have produced a Franklin, than whom no one of the present age has made more important discoveries, nor has enriched philosophy with more, or more ingenious solutions of the phaenomena of nature. We have supposed Mr. Rittenhouse second to no astronomer living: that in genius he must be the first, because he is self-taught."[31] (Another author could have noted that at the moment in history when a committee was needed to draft a declaration of independence and justify a war with Britain, the Continental Congress could call on Jefferson, Franklin, Adams, Livingston, and Roger Sherman.)

This spirited, even angry defense by Jefferson was applauded by most of his readers, but not all. Jefferson had not cited the accomplishments of any New Englanders, and thirty years after *Notes* was published, the omission still smarted for John Adams. He wrote to Jefferson: "Rittenhouse was a virtuous and amiable man an exquisite mechanician, master of the astronomy known in his time . . . But we have had a Winthrop, an Andrew Oliver, a Willard, a Webber, his equals, and have a Bowditch his superior in all these particulars." "But," he continued bitterly, "you know Philadelphia is the heart, the censorium, the pineal gland of the United States."[32]

An important issue in connection with Jefferson's impassioned book concerns the timing of its publication. By the time Jefferson came to write out the longer version of *Notes,* Buffon had already retracted one of his key charges—that North American native peoples were enfeebled and cowardly. In his essay "Sur les Américains" of 1777 he also denied that Europeans in the New World degenerated like their cattle.[33] Raynal had also changed his mind on some topics; on the speech of Chief Logan, for example, he wrote, "Que cela est beau! Comme cela est simple, énergetique et touchant."[34]

The source of Buffon's reversal was none other than Benjamin Franklin, who was a constant friend of Buffon's during his Paris years. In 1751, Franklin had written a important treatise on population (published in 1755) in which he showed that the population of North America had the capacity to grow with a doubling rate every twenty-five years.[35] This was considerably faster than any population growth in Europe and, coupled with America's vast natural resources, presented, as Franklin clearly indicated, a major threat to European hegemony. (In this work, Franklin preceded the British mathematician and population theorist Thomas Robert Malthus by forty years, as Malthus himself acknowledged in his famous 1798 "Essay on Population.")

Many contemporary readers of Buffon complained of his slick changes of position. Nowhere is Buffon's sanctimoniousness more apparent than in this volte-face. Buffon now declared that "in a country where the Europeans multiply so promptly . . . it is scarcely possible that the men degenerate." ("Dans un pays ou des Européens multiplient si promptement . . . il n'est guère possible que les hommes dégénerent.") Apparently without any sense of irony, he now blandly criticized de Pauw for having said, in his 1770 *Récherches Philosophiques sur les Américains,* that American Indians were feeble, beardless and frigid of temperament ("en général tous les Américains, qoique lègers & agiles a la course, étoient destitués de force, qu'ils succomboient sous le moindre sardeau, que l'humidité de leur constitution est cause qu'ile n'ont point de barbe, & qu'ils ne sont chauves que parce qu'ils ont le temperament froid"). But de Pauw had taken these ideas and many of the identical phrases from none other than Buffon himself. Buffon now smugly proclaimed that one cannot ignore the fact that "les Caribes, les Iroquois, les Hurons, les Floridiens, Les Mexicains, les Tlascalteques, les Peruviens, &c," were prevented by the superiority of European arms from demonstrating their full sensibilities, courage, and strength.[36]

Jefferson had a subscription to Buffon's *Histoire Naturelle.* We know that by late 1783, Jefferson had read "Époques de la Nature," in the supplementary volume of 1778 because he corresponded then with James Madison about Buffon's theories.[37] In that case, he must surely have read the previous volume and thus Buffon's retraction. In the 1787 edition of *Notes,* he did add a footnote about Raynal's limited retraction.[38] But neither in the published text nor in later marginal notes on his personal copy did Jefferson acknowledge Buffon's change of position or temper his arguments against, and criticism of, the French scholar.

One could argue that Jefferson had, after all, missed Buffon's essay "Sur les Américains," or that he had not had time to change the manuscript of *Notes*. But Buffon had not retracted any part of his opinions on American wildlife, nor his denigration of the peoples of South and Central America. Moreover, Jefferson was not just arguing against Buffon. He had also to counter those like Robertson and Goldsmith who were copying Buffon's calumnies and who also conveniently ignored his retractions. Once again, Jefferson was stubborn. In *Notes* he had the bit between his teeth; he was not going to be deflected from his mission. So he changed nothing. Science, in this case, had become for him a political weapon.

CHAPTER TEN

Natural History, Slavery, and Race

THOMAS JEFFERSON'S ATTITUDES TOWARD African Americans and the institution of slavery have always excited and inflamed the passions of historians, reformers, and the general mass of decent people. In a typically infuriating set of contradictions, Jefferson detested slavery, but despite his impassioned condemnation of slavery in *Notes on the State of Virginia* and his *Autobiography,* he owned and used slaves. He did not free his own slaves (at least the majority of them). He adamantly rejected the notion of a general emancipation and, equally conspicuously, failed to find effective alternatives. And still, all his life, he condemned slavery and warned of dire consequences should the problem not be solved.[1] Many of these contradictions stemmed from Jefferson's "scientific" estimation of the character and intelligence of black people, whom he considered to be "a race lower in the scale of being" than other races, while he considered American Indians to be part of "us."

Given that *Notes* was his scientific and political manifesto, it would have been surprising if he had not used it to deal head-on with the most contentious issue of his day. Slavery and race together constituted a difficult and, in the end, unresolved issue for Jefferson as a person, as a farmer, as a lawmaker, and as an intellectual. As his writings show, Jefferson was fascinated by racial differences among people, particularly whether the physical and intellectual characteristics of "negroes" were fixed or susceptible to change. The answers had huge potential consequences for his views on human nature and the American body politic.

Science, in this case, however, instead of providing him with answers only embedded him ever more firmly in contradiction. On the subject of race and the institution of slavery, Jefferson remains a paradox even now—perhaps especially now. All these years later and fully a century and a half after Lincoln's Emancipation Proclamation, our confusion and outrage over Jefferson's attitudes toward blacks and his inability or unwillingness to grasp the nettle of emancipation are only magnified when we see the vehemence with which Jefferson the slave owner actually abhorred slavery.

"The whole commerce between master and slave is a perpetual exercise of the most boisterous passions, the most unremitting despotism on the one part, and degrading submissions of the other . . . The man must be a prodigy who can retain his manners and morals undepraved by such circumstances. And with what execration should the statesman be loaded, who permitting one half of the citizens thus to trample on the rights of the other . . . And can the liberties of a nation be thought secure when we have removed their only firm basis, a conviction that in the minds of the people that these liberties are the gift of God? That they are not to be violated but with His wrath? Indeed I tremble for my country when I reflect that God is just; that his justice cannot sleep forever."[2] This dramatic, apocalyptic statement, frozen in time in Jefferson's *Notes on the State of Virginia,* reads no less poignantly today than when Jefferson wrote it more than two hundred years ago. It encapsulates an American dilemma and an American tragedy—both for the man Jefferson and for his new young nation.

Jefferson rightly worried that this biting public condemnation of slavery would earn him enemies. Immediately after the book was published, Jefferson even considered not distributing it. "My reason is that I fear the terms in which I speak of slavery and of our constitution may produce an irritation which will revolt the minds of our countrymen against reformation in these two articles, and thus do more harm than good." He planned to send it out only if the Reverend James Madison approved of it, and he would "have then enough copies to give one to each of the young men at the college, and to my friends in the country."[3]

On reading Jefferson's words for the first time, his friend Charles Thomson wrote: "It grieves me to the soul that there should be such just grounds for your apprehensions respecting the irritation that will be produced in the southern states by what you have said of slavery. However I would not have you discouraged. This is a cancer that we must get rid of. It is a blot in our character that must be wiped out. If it cannot be done by religion, reason and philosophy, confident I am that it will one day be by blood."[4] But he warned Jefferson to remove anything that would seem tolerant of slavery, such as the paragraphs in which he showed his apparent approval of the practices of the Romans.[5]

In view of his opposition to slavery, Jefferson's writings about the African American people (almost all enslaved), who, as he pointed out, made up approximately half of the population of Virginia in 1782, not only inform us of his resolutely negative opinion of black people; they also illustrate a significant aspect of Jefferson's mind. Jefferson was used to thinking in terms of absolutes and, like any natural philosopher, was concerned to find fundamental truths and causes. Jefferson's commitment to the value of natural philosophy led him constantly to look for "truths" upon which to base a personal and political philosophy. The trouble was that several of his most deeply held convictions ("truths") came into conflict, drawing him into irresolvable contradictions. Thus it was with respect to slavery and race. Even though Jefferson abhorred slavery, he held with equal conviction the idea that black people were inferior to whites and that enslaved people could never be incorporated into American society. As a plantation owner with no financial capital, he could not afford to be without slaves, while, paradoxically, he could scarcely afford to keep them. He fervently wished for the abolition of an institution that he could not see a way to end, even within his own household.[6]

Jefferson's fascination with questions of race was both philosophical and practical. It can come as no surprise that many natural philosophers of the eighteenth century thought about the question of race, Buffon and Jefferson being no exceptions. Explorations of the globe had revealed a potpourri of different peoples, about whom the attitudes of Europeans varied from romantic admiration to revulsion. Their character and customs inspired a range of reactions from abstract curiosity to colonial subjugation and exploitation. As knowledge of world geography and the remarkable variety of human beings, often referred to simply as the yellow, red, white, and black peoples, had grown, the existence of different races presented a complex scientific puzzle. The natural philosopher asked: Why are there so many races, and how and why do they differ in the ways that they do? What is the relation of their differences to the places and environments in which they live? They asked one question above all the rest, which they tried to answer both scientifically and theologically: How many species of humans are there? Are some or all of the "races" of humans actually separate species?

Many authors—including Jefferson—skirted around this critical question, not wishing to express a firm opinion. On the one hand, nothing in the Bible story of Creation suggested that God had made more than one kind of human. Some, like the British philosopher Lord Kames, however, had concluded firmly that the diversity of humanity was such that there were several species: if "all men [were] of one species, there never could have existed, without a miracle, different kinds, such as exist at present."[7] Kames was a racialist, but not a racist. He was active, for example, in banning slavery in his home country, Scotland. On the other hand, if all the races *did* compose but a single species, that would create a special series of obligations for the white races with respect to relations with the others. As the kneeling, chained African man asked in the abolitionist image popularized by Josiah Wedgwood, "Am I not a man and a brother?"

The most visible contemporary example of whites behaving in a very nonfraternal way toward blacks in the eighteenth century was the institution of slavery in the New World, but a great deal of slavery existed in the Old World, too. Ever since the Portuguese had explored the west coast of Africa in the sixteenth century, Europeans had created, promoted, and profited from the slave trade, although this did not prevent them from taking a superior attitude toward the slave-owning classes in the New World. Slavery was common in Middle Eastern societies. It had been a commonplace of the civilizations to which Europeans (and men like Jefferson) looked back most reverently: classical Greece and Rome.

There is a curious opposition between Jefferson's and Buffon's views of American Indians and of "negroes." While Buffon claimed that American Indians were a weak and degenerate people, Jefferson found them instead to be honest, brave, and strong, if uncultured. And while Buffon had sympathy for black peoples and hope for their future, Jefferson had little or none.

Buffon had already shown himself to be a particularly weak vessel upon which to place any confidence in the troubled seas of racial identity and characteristics. Even in as cosmopolitan a city as Paris, he had very limited direct experience of different races. In *Histoire Naturelle,* his long chapter on the varieties of the human species drew heavily, and often uncritically, on the accounts of world explorers for descriptions of different races and their distribution by geography. It is also clear that he was strongly influenced by the condition of blacks in slavery.

Buffon wrote: "If the negroes are deficient in genius, they are by no means so in their feelings; they are cheerful or melancholy, laborious or inactive, friendly or hostile, according to the manner in which they are treated. If properly fed, and well treated, they are contented, joyous, obliging, and on their very countenance we may read the satisfaction of their soul. If hardly dealt with their spirits forsake them, they droop with sorrow, and will die of melancholy. They are alike impressed with injuries and favours."

He added, in genuine sympathy, "To their children, friends, and countrymen, they are naturally compassionate; the little they have they cheerfully distribute among those who are in necessity, though otherwise than from that necessity they have not the smallest knowledge of them. That they have excellent hearts is evident, and in having those they have the seeds of every virtue. Their sufferings demand a tear. Are they not sufficiently wretched in being reduced to a state of slavery; in being obliged always to work without reaping the smallest fruits of their labour, without being abused, buffeted, and treated like brutes? Humanity revolts at those oppressions, which nothing but thirst of gold could ever have introduced, and which would still, perhaps, produce an aggravated repetition, did not the law prescribe limits to the brutality of the master, and to the misery of his slave. Negroes are compelled to labour; and yet of the coarsest food they are sparingly supplied."[8]

Remarkably, in setting out his own opinion of blacks in *Notes on the*

State of Virginia, Jefferson used exactly the same dismissive tone and even the same phraseology as Buffon had used to denigrate the Indians. He wrote of African Americans and of blacks in general: "They are more ardent after their female [than whites]: but love seems with them to be more an eager desire, than a tender delicate mixture of sentiment and sensation. Their griefs are transient . . . In general, their existence appears to participate more of sensation than reflection . . . Comparing them by their faculties of memory, reason, and imagination, it appears to me, that in memory they are equal to the whites; in reason much inferior, as I think one could scarcely be found capable of tracing and comprehending the investigations of Euclid; and that in imagination they are dull, tasteless, and anomalous."[9]

This is a hard and unsympathetic set of statements for a man who in most respects was capable of looking beneath the surface of a situation to find its deeper truths. Jefferson admitted, "It will be right to make great allowances for the difference of condition, of education, of conversation, of the sphere in which they [blacks] move." But he reinforced his inherent coldness: "Most of them indeed have been confined to tillage, to their own homes, and their own society yet many have been so situated, that they might have availed themselves of the conversation of their masters; many have been brought up to the handicraft arts, and from that circumstance have always been associated with the whites." Some, he thought, should have profited from this, but "never yet could I find that a black had uttered a thought above the level of plain narration; never see even an elementary trait of painting or sculpture. In music they are more generally gifted than the whites with accurate ears for tune and time, and they have been found capable of imagining a small catch Whether they will be equal to the composition of a more extensive run of melody, or of complicated harmony, is yet to be proved."

Jefferson did not think much of their literary efforts. He devoted nearly a full page of *Notes* to demeaning the writing of Ignatius Sancho, a free black man living in Britain whose patron was the Duchess of Montagu. He wrote plays and a book on music theory; his letters were published posthumously in two volumes in 1782. Jefferson's opinion might have been affected by Sancho's allegiance to Britain: he was a loyal monarchist and vocal supporter in the press of the British side in the Revolutionary War. As for poetry, Jefferson wrote: "Among the blacks is misery enough, God knows, but no poetry." He dismissed the best-known black American poet with the withering words: "Religion indeed has produced a Phyllis Whately; but it could not produce a poet. The compositions published under her name are below the dignity of criticism."[10] Phillis Wheatley, born in Senegal, enslaved at the age of seven, and brought up in a Boston family, was the first published African American poet.

Jefferson was not alone in these views. None other than the Scottish philosopher David Hume wrote an addendum to his *Essays, Moral and Political* (edition of 1742) to the effect that "the negroes and in general all the other species of men" are "naturally inferior to the whites . . . No ingenious manufactures amongst them, no arts, no sciences." And in the influential *History of Jamaica* (1774), James Long described "negroes" as "marked with the same bestial manners." They were stupid and "distinguished from the rest of mankind, not in person only, but in possessing, in abstract, every species of inherent turpitudes."

European philosophers like Buffon, Voltaire, Pierre Louis Maupertius, and many others in the eighteenth century were fascinated by the differences among the races. Beyond conjecture about the geographical basis of racial difference, there were the medico-physiological questions. Whatever the origin of racial differences, what was the manner of their expression? Wherein the skin did color reside, and by what process was it maintained? What was the relationship between skin color and such other characteristics as facial features and the quality of the hair? In one view, for example, it was supposed that black color was produced by the bile, with blacks having a different, darker-colored bile than whites. Another common idea was that black people had black blood. Alternatively, was the cause environmental?

Theories of difference abounded. It was possible that a search for an environmental cause of racial differences—even though that was fashionable in the Age of Enlightenment—was misplaced. Perhaps God, for his own reasons, had made the races just as they are and had placed them where they now live. Another theory was that the indigenous peoples of the

New World were lighter in skin because they represented one of the lost tribes of Israel.[11]

In *Histoire Naturelle,* Buffon firmly supported the view that all humans were one species, the races having become modified according to different environmental conditions in a process that had continued for so many generations that it was not reversible. Color was maintained even when people moved to live in a different environment for several generations (which was the extent of recorded observational experience so far). Following the most commonly held opinion on human diversity, Buffon hypothesized that there were "three causes as jointly productive of the varieties which we have remarked in the different nations of the earth." "First, the influence of the climate; secondly, the food; and thirdly, the manners; the two last having great dependence on the former." Thus the dark skin and characteristic features of Africans were due to generation after generation having grown up under the hot sun, eating a particular diet, and following a particular lifestyle. Of course, no one questioned the premise that a white skin color was both superior and original for the human species.

There were major obstacles to the simple theory that hot climates produced "negroid" (the word means "black") skin color and associated characteristics, such as tightly curled hair (with a flattened cross-section), thick lips, broad nose, and so on. In the first place, black people were not all the same; a member of a North African Nilotic tribe—straight nose, thin lips—and a West African Bantu were hugely different. Intriguingly, however, there was no indigenous "black" race in the New World corresponding to that of the Old World. Buffon explained this partially in terms of the peoples of the New and Old Worlds having had very different histories. If the New World really was newer, as Buffon believed, then perhaps there had not been time for fully black peoples to emerge. He was also convinced that the climate in the New World was inferior, even with respect to the climate of the tropics. "In the New Continent, the temperature of the different climates is more equal than in the Ancient Continent . . . The Torrid Zone is not so hot in America as in Africa . . . Whether we suppose, therefore, the inhabitants of America to be been anciently or recently established in that country, we ought not to find black men there; because their Torrid Zone is a temperate climate."[12]

Buffon's environmental theory might explain why the northern peoples of India are on average lighter skinned than those of the south, and in Europe itself, why northern Italians are usually lighter skinned than those of the south. And it escaped no one's attention that Europeans "left out in the sun," as it were, became tanned. This raised the challenging possibility that different skin colors and their associated characteristics represented grades of a single biological phenomenon—and even that "blackness" might be reversible. Social reformers would rejoice in that case, but there was also the parallel question: If blackness was reversible, could white be changed permanently to black if Europeans lived long enough in hot climates?

In 1787 the Reverend Samuel Stanhope Smith, the president of the College of New Jersey (Princeton), wrote a book in which he elaborated on Buffon's biogeographical treatment of races worldwide.[13] He severely criticized Jefferson for his view of the unredeemable inferiority of blacks and came resolutely to the conclusion that humans make up one species that has become differentiated by climate and the "state of society (diet, clothing, lodging, manners, government, arts, religion, agricultural improvements, commercial pursuits, habits of thinking, and ideas of all kind arising out of this state)."

Indeed, Jefferson was unrelenting in his disparagement of blacks and his emphasis on the features that set them apart. He felt that he was supported by science. "Besides those of colour, figure, and hair, there are other physical distinctions proving a difference of race. They have less hair on the face and body. They secrete less by the kidnies, and more by the glands of the skin, which gives them a very strong and disagreeable odour. This greater degree of transpiration renders them more tolerant of heat, and less so of cold, than the whites. Perhaps too a difference of structure in the pulmonary apparatus . . . They seem to require less sleep." And so on.[14]

Jefferson the natural philosopher was unsure on the question of one species or many, and, interestingly, the physiological questions were only of secondary import to him. "Whether the black of the negro resides in the reticular membrane between the skin and scarf-skin, or in the scarf-skin itself; whether it proceeds from the colour of the blood, the colour of the bile, or from that of some other secretion, the difference is fixed in nature, and is as real as if its seat and cause were better known to us."[15] Instead, he constantly emphasized the sociological and political issues that grew from the differences between black and white, insisting that those differences made any integration of blacks, freely and fully into American society impossible, whatever the causes of the "negroid" condition.

Amid all the swirling issues, color itself was a banner issue for Jefferson. In a passage that presaged Charles Darwin's theory of sexual selection (published a century later) he opined in *Notes* that the importance of color was that it formed the basis of ideas of beauty that would necessarily keep the races apart, each preferring its own members. "Is it not the foundation of a greater or less share of beauty in the two races? Are not the fine mixtures of red and white, the expressions of every passion by greater or less suffusions of colour in the one, preferable to that eternal monotony, which reigns in the countenances, that immoveable veil of black which covers all the emotions of the other race? . . . The circumstance of superior beauty, is thought worthy attention in the propagation of our horses, dogs, and other domestic animals; why not in that of man?"

This was a new and sophisticated idea, but for the lawyer-philosopher it was not enough to state an opinion; Jefferson had to back it up with an argument and evidence. The result was one of the more damning passages in the whole of Jefferson's writings. "Add to these, flowing hair, a more elegant symmetry of form, their own judgment in favour of the whites, declared by their preference of them, as uniformly as is the preference of the Oran-ootan for the black women over those of his own species."[16] This introduction of the "oran-ootan" into a discussion of blacks is, to us (as it was to many of Jefferson's contemporaries), remarkably offensive. It demeaned black women by assuming them to be attractive to apes and essentially lowered all blacks to the level of apes while at the same time reinforcing the superiority of whites.

The titillating possibility of "Natives" being debauched by apes was, however, a popular element in travelers' tales of the eighteenth century. Jefferson may well have read of it in accounts of travels in America and the

East, but he would certainly have had a full exposure to the idea from reading Buffon, who devoted two pages to the subject in his chapter entitled "The Orang-Ooutangs, or the Pongo and Jocko." Buffon repeated stories that orangutans were "passionately fond of women, who cannot pass through the woods, without being suddenly attacked and ravished by these apes."[17] In the 1780s, sub-Saharan Africa and its peoples were still largely unexplored. Apes were little known. Accounts by explorers were the only sources of information, and, as was the case for information about the Far West of North America, explorers often took their information secondhand and exaggerated it wildly to enhance sales of their books. Such accounts soon turned out to be unreliable, but at the time, they presumably appeared authentic.

It is in the nature of science and scientists to believe that for every problem there is a solution and that experimentation is one key to finding that solution. In 1787, during a tour of France and Italy, Jefferson fell into conversation with the American physician Edward Bancroft (an associate of Franklin's and apparent loyalist who later turned out to have been a double agent in the Revolutionary War) on the subject of assimilating slaves into society.[18] The subject was some experiments made by Quaker landholders in Virginia to wean slaves from slavery by progressively emancipating them through first giving them access to their own land, on which they would be sharecroppers.[19] This was allowed in Virginia by the Act of 1782, which stipulated that "slaves so set free, not being in the judgment of the court, of sound mind and body, or being above the age of forty-five years, or being males under the age of twenty-one, or females under the age of eighteen years, shall respectively be supported and maintained by the person so liberating them, or by his or her estate."[20]

Bancroft later wrote from London asking for more details for "friends of mine, who are warmly and benevolently active in the Society abolished here for abolishing negro Slavery." He remembered that Jefferson had been negative about the scheme, for "after a trial of some time it was found that Slavery had rendered them incapable of self-government, or at least that no regard for futurity could operate on their minds with sufficient force to engage them to anything like constant industry or even so much of it as would provide them with food and Cloathing and that the most sensible of them desired to return to their former state."[21]

In view of everything else that Jefferson wrote and thought about slavery, his reply to Bancroft is remarkable. He alluded to the various experiments that Quaker landholders in Virginia had already tried, and also the experience of "a young man . . . (Mr. Mayo) who died and gave freedom to all his slaves." Although Jefferson thought the results were discouraging, he told Bancroft that when he (Jefferson) returned to America, "I shall endavour to import as many Germans as I have grown slaves. I will settle them and my slaves, on farms of 50. acres each, intermingled, and place all on the footing of the Meteyares [French sharecroppers] of Europe. Their children shall be brought up, as others are, in habits of property and foresight, and I have no doubt but that they will be good citizens. Some of their fathers will be so: others I suppose will need government. With these, all that can be done is to oblige them to labour as the labouring poor of Europe do, and to apply to their comfortable subsistence the produce of their labour, retaining such a moderate portion of it as may be a just equivalent for the use of the lands they labour and the stocks and necessary advances."[22]

This experiment comparing two uneducated laboring classes was never tried. Any inclination that Jefferson might have had in this direction was overwhelmed by the practical difficulties, but a more fundamental obstacle will have been Jefferson's constant negative opinion of the possibilities of improvement within the enslaved people; blacks, as a lesser race, would never attain the qualities necessary for citizenship.

So Jefferson, less and less optimistically, stuck with his first solution to the problem of slavery. It was as simple as it was essentially impossible. For Jefferson, emancipation could be contemplated only if it was coupled with repatriation to Africa or to emigration to colonies in the West Indies.[23] He introduced a bill for the Virginia Legislature to this effect in 1779; he repeated the plan in *Notes,* and he was still promoting it at his death.

Jefferson proposed "to emancipate all slaves born after passing the act . . . They should continue with their parents to a certain age, then be brought up, at the public expence, to tillage, arts or sciences, according to their geniuses, till the females should be eighteen, and the males twenty-one years of age, when they should be colonized to such place as the circumstances of the time should render most proper, sending them out with arms, implements of household and of the handicraft arts, feeds, pairs of the useful domestic animals, &c. to declare them a free and independant people, and extend to them our alliance and protection, till they shall have acquired strength; and to send vessels at the same time to other parts of the world for an equal number of white inhabitants; to induce whom to migrate hither, proper encouragements were to be proposed."[24]

This scheme was typically complicated; if put into effect, it would have steadily reduced the extant black population of the United States to fewer and fewer, older and older people. Jefferson never gave up on the idea; he repeatedly tried to find new ways to put it into practice in different host countries. But, not surprisingly, any scheme to siphon off the younger generation of American-born blacks to colonies in Africa or the Caribbean ran headlong against yet another of Jefferson's strongly held opinions: that it was morally wrong to break up black families.

Yet another pressing difficulty with Jefferson's "colonization" (repatriation) scheme was that removal of the slaves would leave a huge gap in the agricultural labor pool, and that would have to be met by immigration of whites. In principle, Jefferson was against this as well. In a separate section of *Notes* he warned against the dangers of immigration proceeding unchecked. The potential result was that America would be flooded with immigrants who lacked education in American values. They would "bring with them the principles of the governments they leave . . . or, if able to throw them off, it will be in exchange for an unbounded licentiousness, passing, as is usual, from one extreme to another." He summed up his disapprobation of European immigrants: "In Europe the object is to make the most of their land, labour being abundant; here it is to make the most of our labour, land being abundant."

As much as he hated slavery, Jefferson could not accept any plan for emancipation that would have allowed blacks to live freely with white citizens of the United States. He never changed his mind on this. As he wrote in his *Autobiography*, "Nothing is more certainly written in the book of fates, than that these people are to be free; nor is it less certain that the two races, equally free, cannot live in the same government. Nature, habit, opinion have drawn indelible lines of distinction between them."[25] By "opinion," Jefferson seems to have meant that he did not think blacks were constitutionally worthy of equality with whites.

There were further reasons for racial incompatibility. "Deep rooted prejudices entertained by the whites; ten thousand recollections, by the blacks, of the injuries they have sustained; new provocations; the real distinctions which nature has made; and many other circumstances, will divide us into parties, and produce convulsions which will probably never end but in the extermination of the one or the other race.—To these objections, which are political, may be added others, which are physical and moral." Jefferson undoubtedly feared that a large population of blacks, newly released from unjust, involuntary, and often inexpressibly harsh servitude would constitute a civil danger. After 1791 the example of the successful Haitian slave rebellion was before all Americans, and local outbreaks, such as Gabriel's Rebellion of 1800, underlined the threat.[26]

Although there were precedents for significant manumission by individual owners—Robert Carter of Virginia freed more than four hundred of his, beginning in 1791[27]—any scheme of national emancipation would necessarily have immense political, economic, as well as social repercussions. Jefferson may have been loath to support any plan that was coercive on slave owners; any such plan would have destroyed his political career.[28] In this way, Jefferson's philosophical positions, each perfectly logical, came into direct conflict not only with one another but with practicality. Finally, repatriation would never have been possible because, if nothing else, the cost was prohibitive. Although Jefferson kept trying, he reconciled himself to the fact that emancipation would not occur in his lifetime.

Jefferson's views on race and slavery did not change with the passing years despite all the criticism that was heaped upon him. Meanwhile, black Americans were, to use Jefferson's own words, "gaining daily in the opinions of nations, and hopeful advances are making towards their re-establishment on an equal footing with the other colors of the human family." When the French Jesuit priest and revolutionary Henri Gregoire sent Jefferson a book extolling the "Literature of Negroes," Jefferson wrote back: "Be assured that no person living wishes more sincerely than I do, to see a complete refutation of the doubts I have myself entertained and expressed on the grade of understanding allotted to them by nature, and to find that in this respect they are on a par with ourselves." He added the interesting observation that "whatever be their degree of talent it is no measure of their rights."[29]

One of the more accomplished contemporary African Americans was Benjamin Banneker, a free man who, ironically, was interested in many of the subjects that fascinated Jefferson. He was a surveyor who assisted Pierre-Charles L'Enfant and Andrew Ellicott in laying out the street plan for the nation's new capital. Beyond that, he had used his mathematical skills to devise an almanac of the heavens, which he sent a copy of to Jefferson. The extent of Banneker's originality and genius is still unclear, but at the time he was rapidly becoming a cause célèbre in the eyes of abolitionists, and Jefferson came under pressure to acknowledge him.

Jefferson replied graciously to Banneker's gift. "Nobody wishes more than I do to see such proofs as you exhibit, that nature has given to our black brethren, talents equal to those of the other colors of man, and that the appearance of want in them is owing mostly to the degraded condition of their existence, both in Africa and America."[30] This graciousness should not surprise us. Jefferson was, after all, the man—in legend, at least—who had been criticized for raising his hat to a black man he passed on the street.[31]

However, at the same time as he wrote to Banneker, he wrote quite differently about Banneker to his friend Joel Barlow: "We know he had spherical trigonometry enough to make almanacs, but not without suspicion of aid from Ellicot, who was his neighbour and friend, and never missed an opportunity of puffing him. I have a long letter from Banneker, which shows him to have had a mind of very common stature indeed."[32] Jefferson also told Barlow that he had written Gregoire "a very soft answer": "It was impossible for doubt to have been more tenderly or hesitatingly expressed than that was in the Notes of Virginia, and nothing was or is farther from my intentions, than to enlist myself as the champion of a fixed opinion."

Five years later, in 1814, with the mixed success of the British experiments for a colony in Sierra Leone (to which they sent many of the blacks who had fought on the loyalist side in the Revolutionary War) and a brutally successful slave revolt in Haiti, Jefferson still hoped that his scheme of repatriating educated young black people might work, but he was not optimistic. He wrote to Edward Coles that whatever the extenuating conditions of their existence, blacks were "pests in society by their idleness, and the depredations to which this leads them. Their amalgamation with the other colour produces a degredation."[33]

CHAPTER ELEVEN

The Color of Their Skin

JEFFERSON CONSISTENTLY WROTE THAT the physical, intellectual, and even moral differences between the black and white races were fixed in nature and were, as a result, irreconcilable. He did not agree with the view, common among European authors, that color was environmental in origin and therefore modifiable. And nowhere in his discussion of blacks as blacks did Jefferson refer to what was obvious to many—that the differences between the races readily became blurred. That fact was more than obvious from the growing number of Virginians of mixed parentage and a range of physical attributes. Many of his male Virginian slave-owning relatives and neighbors took black concubines or simply forced themselves on their black slaves. Jefferson's own slave "family," as he referred to the domestic slaves at Monticello, included many of mixed race. Foremost among these was Sally Hemings, the half-sister of Jefferson's late wife (they shared the same father); she was an "octoroon"—three of her four grandparents having been white.[1] Jefferson had been conducting, as it were, his own experiments in racial mixture.

Several of the children that Sally Hemings had by Jefferson easily passed as white. But by Virginia law everyone was a slave whose mother was or had been a slave, regardless of the particular mixture of blood. Those children were still considered inferior by the man who as late as 1814 wrote of black people that (continuing the passage quoted at the end of the previous chapter) "their amalgamation with the other color produces a degradation to which no lover of his country, no lover of excellence in the human character can innocently consent."[2] For Jefferson, clearly, the only thing that mattered was being pure white; any mixture was "lesser," and this applied even to his own offspring.

At one point, Jefferson's friend Charles Thomson, who admired what Jefferson said about the imperative to abolish slavery, suggested that it would be better for him to omit his comments on the inferiority of blacks from *Notes on the State of Virginia.* Instead, Jefferson included a sort of compromise statement that is half self-justification and half retraction. "To our reproach it must be said, that though for a century and a half we have had under our eyes the races of black and of red men, they have never yet been viewed by us as subjects of natural history. advance it therefore as a suspicion only, that the blacks, whether originally a distinct race, or made distinct by time and circumstances, are inferior to the whites in the endowments both of body and mind. It is not against experience to suppose, that different species of the same genus, or varieties of the same species, may possess different qualifications. Will not a lover of natural history then, one who views the gradations in all the races of animals with the eye of philosophy, excuse an effort to keep those in the department of man as distinct as nature has formed them? This unfortunate difference of colour, and perhaps of faculty, is a powerful obstacle to the emancipation of these people."[3]

While Jefferson closed his eyes to the all-too-obvious fact that miscegenation between the races tended to blur their distinctiveness, he had a scientist's fascination with a different situation—one that quickly came into focus in the mid to late eighteenth century. It was becoming obvious that there were naturally occurring examples of an apparent change in skin color from black to white. Once again, Buffon had documented the phenomenon.

As was well known, black families sometimes produced an albino ("white negro") child. Albinism appeared just as frequently among American Indians. Buffon personally examined a full albino "negress" from Dominique named Genevieve, aged eighteen, who was born of two black parents. He described her minutely in *Histoire Naturelle,* reproducing her portrait in voluptuous frontal nudity. This full-page plate of Genevieve, posed with African symbols, a basket of fruit and vegetables, and a European cupboard, was endlessly reproduced over subsequent decades. Buf-

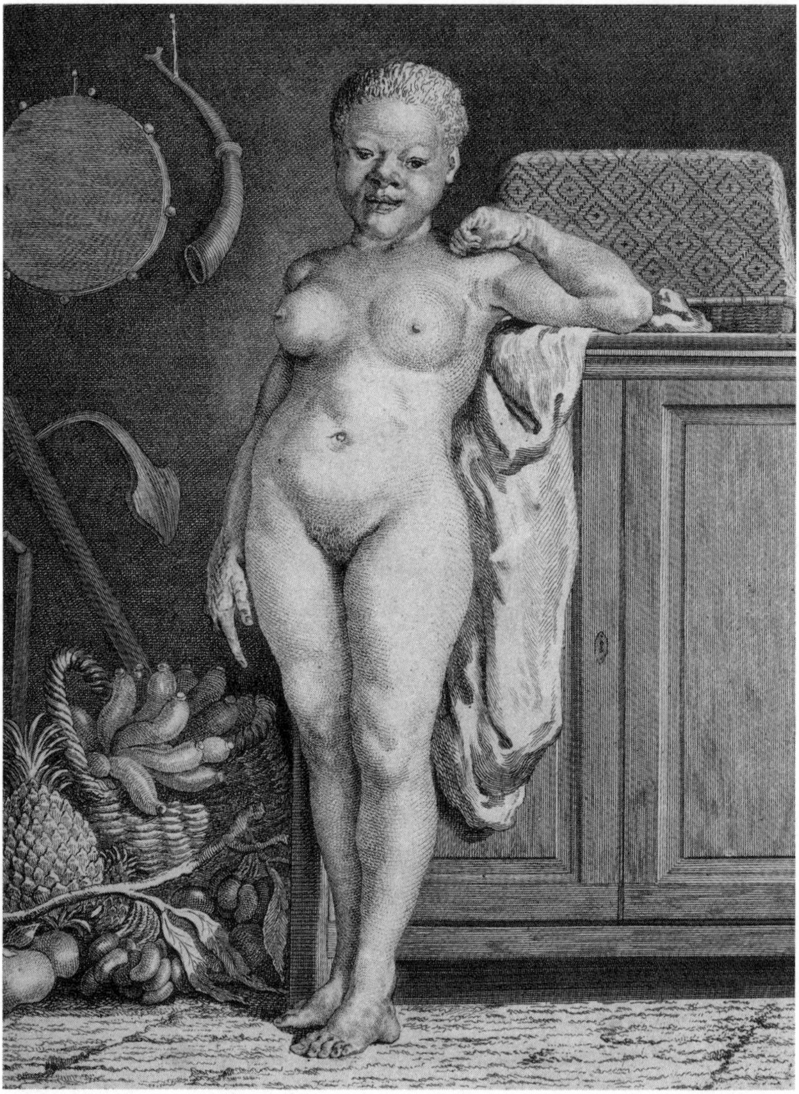

Buffon's engraving of the young albino African woman "Genevieve" from Dominique, published in *Histoire Naturelle,* was widely copied in the eighteenth and nineteenth centuries. Courtesy of the Academy of Natural Sciences. © Ewell Sale Stewart Library/VIREO.

fon was of the opinion that albinism was more common in the New World than the Old and was restricted to people living six or eight degrees north or south of the equator. He concluded that "white Negroes . . . are only Negroes who have degenerated from their race, and not a particular and permanent species of men."[4]

Although Jefferson certainly read about albinism in Buffon's *Histoire Naturelle* and Diderot's *Encyclopédie,* he knew about it personally, too; there were cases of albinism among the slaves of his friends and neighbors. His fascination with albinism was neither prurient nor idle. He explored seriously whether, and under what conditions, skin color and characteristics like facial features might become modified. And he was sufficiently interested in the condition to devote a section of *Notes* to the subject. At the end of chapter 6, having dealt with plants, mammals, and birds, and just before a cursory paragraph on fishes and insects, Jefferson added a note concerning "white negroes" born of parents "who, though black themselves, have in rare instances white children called Albinos."

Perhaps he was determined to improve on what Buffon had written; but he must also have been challenged to find out what this condition meant for his view of the distinctiveness of races. Some authorities thought that albinism was simply (or not so simply) the body trying to revert to its original preferred white condition. To Jefferson, who saw the races as fundamentally separate biologically, albinism was a challenge to his view that they must be kept apart politically.

Late in the drafting of *Notes,* Jefferson evidently wrote to several friends for information on the subject of albino African Americans. His brother-in-law Henry Skipwith wrote to describe for Jefferson three albino sisters born to a black family he owned. Their parents were "the ordinary color of blacks (not jet)," and they had three white children and two black. Two of the sisters later had children by black men, "and each had issue female children of the ordinary complexion of light coloured negroes."[5] In Skipwith's experience, albinos were usually women. He knew, for example, of a "very robust white negro slave," a woman who was born of black parents and who "had issue (by a black man) of a daughter of the jet hue." The only exception he knew was through his neighbor Charles Lee. who owned an "elderly, tall, stout, white negro man slave" whose parents were both black. These were probably all true albinos. They were healthy and active, but all had the distinguishing characteristic of extremely weak eyes that were painful in bright sunlight.

Charles Carter of Fredericksburg reported to Jefferson concerning a "white negro" woman belonging to his father. Her parents "were brought from Guinea and were remarkable for their health and longevity. I think they may be styled dark Mulattoes, of which colour they had three other children." The young woman in question had skin that was "a sallow white with a great many dark spots promiscuously scattered over it somewhat like the freckles on a white person's face but rather larger. It is disputed whether those spots are natural or acquired, but the blacks inform me they are natural . . . Her hair is thick curled and as coarse as the black negroes but differs in colour . . . her senses are as perfect as any person's except her sight which fails her very much when exposed to the sun . . . She has had a child by a true black, which had the complexion of the mother."[6] (One might question the status of this woman as a full albino, although she did have the characteristic feature of weak eyes.)

Jefferson summarized these observations: "The circumstances in which all the individuals agree are these. They are of a pallid cadaverous white, untinged with red, without any coloured spots or seams; their hair of the same kind of white, short, coarse, and curled as is that of the negro; all of them well formed, strong, healthy, perfect in their senses, except that of sight, and born of parents who had no mixture of white blood. Whatever be the cause of the disease in the skin, or in its colouring matter, which produces this change, it seems more incident to the female than male sex."[7]

Beyond albinism, yet another phenomenon tended to dramatically blur the apparent distinctiveness of whites and blacks. In the fourth supplementary volume of *Histoire Naturelle* (1778), Buffon devoted considerable space to the subject of "piebald" blacks, whose skin was marked by large patches of white skin. These patches appeared after birth, sometimes quite late in life, and gradually spread, but never, apparently, make the whole body white. Buffon gave illustrations of two such cases; one was apparently a black child suffering from vitiligo universalis, which produces the piebald condition. (The other was quite different; it was probably a case of the much rarer phenomenon of melanism in an otherwise white child.)[8]

Buffon's example of vitiligo was based, oddly enough, not on a person he had seen, but on the portrait of a young woman from Dominique that had been sent to Europe. In publishing a copy of this portrait of Maria Sabena in *Histoire Naturelle,* within the same essay as his treatment of albinism and the picture of Genevieve, Buffon made her a sensation, too, and public displays of albinism and vitiligo soon became common as new examples were imported from the Americas and displayed in sideshows. People with names like "Piebald Girl," and "Leopard Boy" made a living by charging for a view of their changing spots.

Buffon believed that the piebald condition was due to the crossing of a full black and an albino parent.[9] Jefferson's correspondents informed him differently. Unlike Buffon, Jefferson was personally familiar with vitiligo, as it occurred among the African families enslaved in Virginia. In fact, an example had been recorded by the Virginian plantation owner and explorer William Byrd nearly a hundred years before Jefferson wrote on the subject in *Notes.* At the age of three, an otherwise fully black child, born to black parents, began to develop white spots on the neck and chest. By the time he was eleven (and living in England) most of his torso was "dappel'd with White Spots." These spots were "wonderfully white, at least equal to the Skin of the fairest Lady." And Byrd added a very interesting additional observation: the white patches "are not liable to be Tann'd."[10] A little later, a case was reported to the Royal Society in London of a forty-year-old woman, "originally as dark as the most swarthy African, in whom, over some fifteen years, four parts in five of the skin, starting with the mouth and fingers," had become "white, smooth, and transparent, as in a fair European." The author of the account, "the Reverend and ingenious Mr Williamson of Maryland," rejected the notion that the condition was caused by a condition of the bile.[11]

Jefferson knew of a similar example: "a negro man . . . born black, and of black parents; on whose chin, when a boy, a white spot appeared. This continued to increase till he became a man, by which time it had extended over his chin, lips, one cheek, the under jaw and neck on that side. It is of the Albino white, without any mixture of red, and has for several years been stationary. He is robust and healthy, and the change of colour was not accompanied with any sensible disease, either general or topical." It will be noted that Jefferson was at pains to state that the white skin of these people was not comparable to that of Europeans, lacking any traces of red. Byrd thought otherwise, and his comparison of the skin of his "Negro-Boy" to that of the "fairest Lady" would have irritated Jefferson.

Jefferson was not in Philadelphia for the exhibition by Dr. John Morgan in 1784, at the American Philosophical Society in Philadelphia, of "a motley Coloured, or Pye Negro girl" and a "Pyed Mulatto Boy" from Guadaloupe.[12] Morgan told the society that the "alteration of the natural colour" of the girl Adelaide was directly comparable to that of Buffon's Maria Sabena and that "she had a negro father and negro mother. The body of the girl showed patches of white skin. The boy, named Jean Pierre, was "entirely the colour of a mulatto," except for the white portions; the pied condition also existed in his father and mother. Jefferson was in Philadelphia after the 11th of that month on his way to Paris. A display like that was not something he would have missed, but it might have come too late for him to include in his manuscript for *Notes.*

It seems unlikely that Jefferson, had he been in town on July 23, 1796, would have failed to attend the following spectacle, advertised in a broadside: "A GREAT CURIOSITY. There is a man at present at Mr. LEECH'S Tavern, the sign of the Black Horse, in Market-street, who was born entirely black, and remained so for thirty eight years, after which his natural body colour began to rub off, which has continued till his body has become as white and fair as any white person, except for some small parts, which are changing fast . . . He may be seen . . . at one Quarter of a dollar each person, children half price."[13]

The case of this man, named Henry Moss, was investigated fully by Benjamin Smith Barton, the Philadelphia physician, anthropologist, and botanist. Moss, said Barton, came from "a very mixed breed, if I may be allowed the use of an expressive phrase, which is frequently employed by naturalists when treating of other animals. Moss was a free man who had fought in the Revolutionary War. His father was "very black," although he had had an Indian mother. Moss's own mother was "mullato," and her mother had been an Irishwoman. Moss's vitiligo (which Barton refers to as *Leucaethiopica humana*) began when he was about twenty, first at the base of his fingernails. By 1806 the "greater part" of his body was a "delicate sanguine-white" (not the dead white that Jefferson insisted was typical of albinos). The change progressed only in spring and summer, never in winter. His epidermis did not peel off, and the white skin did not tan, although it did acquire temporary freckles. Cuts on Moss's legs healed to his original chestnut-black color. The hair of his beard and on his chest became straight and fell out easily. His eyesight was not affected.[14]

After *Notes* was published, the Reverend James Madison wrote to tell Jefferson of an Indian with a similar condition who was known to President Stiles of Yale College. The man was fifty years old and, "for near two Years past, has been gradually whitening. It began on his breast, and has transfused itself throughout the body to the Extremities. Above half his hands and Feet and Toes are yet of the Indian colour and his Face pied. The skin on the other parts is become a clear English White *with English Ruddiness.*"[15]

Behind these investigations into the nature of albinos and the occurrence of "piebald" skin conditions, was an important philosophical question. If white was the original color from which other races had diverged owing to climate or some other circumstance, was it possible for "blackness" to be reversed? Putting it crudely, the black condition might be *cured.* Samuel Stanhope Smith, for example, concluded that cases like Henry Moss represented the results of blacks living alongside people of European stock, although he was also sure, even though whites could tan, that it would be impossible for whites to become similarly transformed by local circumstances into blacks.

In 1797, Dr. Benjamin Rush at Philadelphia wrote to Jefferson to say that he was preparing a paper for the American Philosophical Society in which he would argue that "the black Color (as it is called) of the Negroes is the effect of a disease of the Skin in the Leprous kind."[16] Rush, a signer of the Declaration of Independence and a controversial figure during the 1793 yellow fever epidemic in Philadelphia, was totally opposed to slavery and argued that the supposed inferiority of backs was only a product of their circumstances. Typical of Rush is an entry in his Commonplace Book for August 22, 1793: "Attended a dinner . . . to celebrate the raising of the roof of the African Church. About 100 white persons chiefly carpenters dined at one table—who were waited upon by Africans. Afterwards about 50 black people sat down at the same table, who were waited upon by the white people. Never did I see people more happy. Some of them shed tears of joy."[17]

Rush could only find anecdotal evidence for his idea about leprosy—reports of the skin in white people turning black and of the skin of black people turning white. He also repeated the myth that the albino condition was produced by "women being debauched in the woods by the larger baboon, ourang-outang." If correct, Rush's theory would have had a major effect on attitudes toward race because the "negroid" condition would be curable. Rush thought that the case of Henry Moss suggested that perhaps nature "has begun spontaneous cures of this disease in several black people in this country."

There was, however, a sting in the tail of this argument. Even if blackness was caused by a disease that "should entitle them to a double portion of our humanity," white people would have to maintain "that prejudice against such connections with them, as would tend to infest posterity with any portion of their disorder. This may be done . . . without offering violence to humanity, or calling in question the sameness of descent, or natural equality of mankind." Rush's "suppositions" also required, as he wrote to Jefferson, retaining the "existing prejudices against matrimonial connexions with [Negroes]." Thus, for all his humanity, Rush only reinforced the need for separation of the races. Thus, in yet another example of the contradictions in Jefferson's natural philosophy, the two men, one sympathetic to blacks and the other not, were forced to the same conclusions.

PART FOUR

Useful Knowledge

CHAPTER TWELVE

The Paris Years

"IT WAS IN FRANCE, where he resided nearly seven years, and until the revolution had made some progress, that his disposition to theory, and his skepticism in religion, morals, and government, acquired full strength and vigor."[1] Taken out of context, this statement might almost be seen, today at least, as a compliment to Jefferson; coming from a contemporary Federalist opponent in an election race, it was meant nastily, as a condemnation. But it was true. In his Paris years, intellectually, once he had settled in, Jefferson was in his element. Paris—at least its upper crust—offered a heady mix of art, literature, music, philosophy, science, invention, intrigue, and social revolution. The food was delicious, the wines even more so. The streets were not paved with gold, but the banks of the Seine were lined with booksellers' stalls full of enticing titles in a dozen languages.

It was in Paris that Jefferson also encountered a new world of science and inventiveness. In the contemporary sense, the French word "philosophe" fitted Jefferson perfectly. In the Age of Reason, a philosophe was someone who applied his (mostly his) formidable learning and intellectual powers not just to solving abstract problems of moral philosophy or the meaning of life but to addressing the pressing issues of the day, in science, politics, the social state, and the human condition. Philosophes, people like Rousseau, Diderot, and Buffon, wrote extensively and widely. No subject could fail to be improved by a combination of pure reason and practicality. Thomas Paine, for example, author of the political pamphlet *Common Sense* and someone Jefferson admired greatly, "invented an iron bridge which promises to be cheaper by a great deal than stone, and to admit of a much greater arch."[2]

Jefferson arrived in Paris on April 6, 1784, to serve as U.S. commissioner and minister. Peace negotiations with Britain having been completed, the United States needed to set up commercial treaties and arrange loans. John Adams was already there, as was Franklin, whose place Jefferson would take (Jefferson famously remarked that no one could actually *replace* Franklin). Having spent the last year in Congress, where he had drafted much legislation, Jefferson was not awed by his duties. Paris itself was a different matter at first.

To serve his country, Jefferson had to surmount his fundamental opposition to monarchy and all forms of aristocracy. He needed to move into a refined world where elaborate dress, a courtly style and manners, and sophistication (and much affectation) were paramount. While the immensely popular Franklin could get away with presenting himself as a version of an American rustic, Jefferson had to adopt court manners and put aside his usual plain dress for smart frockcoats and ruffled shirts—everything that he normally abominated, as contemporary portraits show.[3] He had to buy a dress sword (for ninety-six francs), plus a sword belt (for three francs), but he must have been careless with the sword, for two years later he had to have it repaired. The next year he bought another one. But he already owned at least one sword; his record books show he had that one repaired in 1783.

In addition to purchasing swords, thermometers, and every other kind of interesting or useful object, Jefferson apparently had a penchant for pistols, purchasing over his lifetime far more than would seem to have been necessary. He was the last man one could imagine engaging in a duel, so he must have bought them for self-defense during his long travels by horseback or carriage. And perhaps also because they were appealing works of craftsmanship.

Paris society was not just froth and furbelows (and dress swords). The work of negotiation was tough and unremitting. On the positive side, Jefferson had now moved to the intellectual capital of the European world. Rousseau and Voltaire were dead, but Jefferson's friends Lafayette and Chastellux soon introduced him to the people he wanted most to meet: the Duc de Rochefoucauld, the Marquis de Condorcet, the Comte de Volney, Pierre

Samuel du Pont de Niemours, and a number of prominent intellectual clerics, such as the Abbés Morellet, Arnauld, Chalut, de Mably, and Barthélemy. And, of course, the Comte de Buffon. (De Volney and du Pont eventually found their way as exiles to America.) He frequented the salons of such formidably influential women as Sophie, Marquise de Condorcet, and Madame de Staël. Madame de Tessé, aunt of Lafayette, was a particular favorite; they had common interests in gardening and began a lifelong correspondence and exchange of plants and seeds. Notwithstanding claims that Jefferson was something of a misogynist, or at least uncomfortable around the opposite sex, he evidently relished the company of educated and attractive women.[4] Throughout his life, his letters to women, and those to Madame de Tessé particularly, are notable for their elaborate compliments, gentle teases, and affectionate style. And it was in Paris that he fell in love with Maria Cosway, an artist and the wife of the British miniaturist Richard Cosway. He had, however, the typical male sexist attitudes of the eighteenth century: he believed that a woman's place was in the home, not in politics.[5]

In Paris, everywhere Jefferson turned there was something new and exciting, glittering and challenging, and not just on the ground. All Paris, it seemed, was looking up to the skies. The flight of heavier-than-air machines, a dream since before the time of Leonardo da Vinci, had been achieved. Ballooning was the new scientific art, fascinating philosophers and the public alike, and on June 4, 1783, the seemingly impossible had been achieved by Jacques-Étienne Montgolfier and his brother, Joseph-Michel, with an unmanned balloon. The new "aerostatic machine" was made of cloth and paper and had a diameter of about one hundred feet. It weighed five hundred pounds. The flight only lasted ten minutes; the fabric had too many joins that let the hot air out.

On September 19, 1783, Louis XVI witnessed a Montgolfier balloon take aloft a sheep, a cock, and a duck. Eight weeks later two men (Jean-François Pilâtre de Rozier, a physicist, and François Laurent, Marquis d'Arlandes) made the first manned flight, covering six and a quarter miles and rising to three thousand feet. Benjamin Franklin and John Adams were present in the crowd at Versailles that day.

The Montgolfier brothers' first balloon took off from Paris on September 19, 1783. From Faujas de Saint-Fond, *Description des Expériences de la Machine Aérostatique* (1784). Courtesy of the American Philosophical Society.

There had been just the same excitement in Philadelphia when Jefferson left. He had arrived in that city just in time to witness the first flight (on May 20, 1784) of a homegrown American balloon—not one big enough (or safe enough) for carrying people, but nonetheless it flew. "I have had the pleasure of seeing three balons here. The largest was 8.f. diameter, and ascended about 300 feet."[6] The balloon's maker was Dr. John Foulke, who, with George Washington's encouragement, had begun to experiment with flying small balloons made of paper.[7] Evidently the prospect of sending these uncontrollable fiery devices up into the air, possibly to land on a house or in a hay field, did not deter any of these early inventors.

The scientific principle of the Montgolfier invention was simple enough and was obvious to Jefferson. Hot air rises, and Robert Boyle, in setting out his gas laws in 1622, had put that commonplace fact into scientific form. Heated air expands in volume and, in the process, becomes less dense (and lighter per unit volume) in direct proportion to its temperature. The simplest way to make a balloon lighter than air was to fill it with heated air. Hot air balloons were dangerous, however, because then, as now, they required a portable source of heat. Lacking convenient propane cylinders, the eighteenth-century balloonist hung a basket beneath the open end of a huge sac made of silk or paper and lit a fire in it. The preferred fuel was dry straw and shredded wool, which generated less smoke than did wood. Balloons were expensive: a small model could be made of paper, but a balloon to raise two men required hundreds of square yards of silk taffeta or cotton, and paper. Except for the fuel source and the fabric, not very much has changed since the time of the Montgolfiers. But although hot-air-balloon flight now is commonplace, then it was almost miraculous, and Jefferson was fascinated by it.

Jefferson saw as one of his duties in Paris, reporting back to the United States everything technical and useful, as well as diplomatic, that he learned. He reported on scientific matters to John Jay, the secretary of state, and to a range of friends, including Francis Hopkinson, Charles Thomson, the Reverend James Madison, Hugh Williamson, David Rittenhouse, Thomas Paine, and Joseph Willard. And Jefferson was soon writing back home about developments in ballooning. Within weeks of arriving in Paris, he sent a detailed accounting, complete with a table, of the flights that had been made, the construction of the balloon, and the fuel used.[8]

Almost from the beginning, two different fuels were used for balloons, hot air (rarefied air) and "inflammable air" (hydrogen gas). The first hydrogen balloon ascent was made from the Camp de Mars in Paris on December 1, 1783, by Jacques Charles and Nicolas-Louis Robert. The English chemist Henry Cavendish had recently discovered a way to generate hydrogen by mixing iron filings and sulfuric acid—not a very much safer concoction to hang in a basket under a balloon than a fire pit. Jefferson later acquired the formula from him.[9]

Jefferson reported the first death of a balloonist, when Pilâtre de Rozier was killed trying to cross the English Channel in a hydrogen balloon coupled to a hot air balloon. Having been "waiting some months at Boulogne for a fair wind to cross the channel," he "at length took his ascent with a companion . . . Being at a height of about 6000 feet, some accident happened to his balloon of inflammable air, it burst, they fell from that height, and were crushed to atoms . . . It is suspected the heat of the Montgolfier rarified too much the inflammable air of the other and occasioned it to burst."[10]

As even his earliest letters home show, Jefferson very quickly analyzed both the positives and the negatives of this first human adventure into a new dimension. It was, of course, a huge achievement. But there were obvious drawbacks, starting with the lack of steering, even though fanciful attempts were made to use sweeps and sails. Jefferson summarized the pros and cons. Balloons could be used for "traversing deserts, countries possessed by an enemy, or ravaged by infectious disorders, pathless & inaccessible mountains. conveying intelligence into a beseiged place, or perhaps enterprising on it, reconnoitring an army &c. throwing new lights on the thermometer, barometer, hygrometer, rain, snow, hail, wind & other phenomena of which the Atmosphere is the theatre. the discovery of the pole which is but one day's journey in a baloon. from where the ice has hitherto stopped adventurers. raising weights; lightening ships over bars. housebreaking, smuggling &c. some of these objects are ludicrous, others serious, important & probable."[11] In 1785, Jean-Pierre Blanchard and John

Jeffries (an American) successfully crossed the English Channel in a hydrogen balloon, and suddenly the balloon was not just a novelty or a toy, but a serious matter. Natural barriers and national boundaries were invisible to it. If a balloon could be devised to carry more than a couple of passengers, it could become a serious military threat.

The first successful manned flight in America was made in Philadelphia by Jean-Pierre Blanchard, the man who had first flown the English Channel. The Philadelphia flight occurred in 1793, and both Jefferson and Washington were there to observe it. Jefferson's friend Dr. Caspar Wistar and other colleagues used the occasion for an experiment. Wistar asked Jefferson to persuade Blanchard to take up some vials of water. At altitude the vials were to be emptied and then capped so that Wistar could analyze any differences in the air from that at ground level.

We know most about Jefferson's interests in Paris from his correspondence and his purchases. Unfortunately he did not share with his American or British friends details of the more intellectual side of his stay. He sought out Condorcet early in his visit, for example, and they talked a lot of politics and discussed subjects like European attitudes toward slavery. He carefully translated and copied out Condorcet's notes on that.[12] But nowhere did he record what he and the illustrious mathematician talked about on scientific subjects.

Jefferson made many reports on the technological innovations he saw in Europe. His lifelong interest in the practical application of science to farming and industry, astronomy and cooking, and everything in between, is now legendary, but it is still possible for an item in the extraordinary list of his interests to bring one up short. For example, Jefferson had an idea for a steam engine. Having seen large steam engines at work, he thought "it might be possible to economize the steam of a common pot, kept boiling on the kitchen fire until its accumulation should be sufficient to give a stroke." There might even be "enough in a day to raise from an adjacent well the water necessary for daily use."[13] The idea seems impractical and was not mentioned again.[14]

The invention of steam power and its practical application to industry and navigation was one of the great breakthroughs in the late eighteenth century. The steam engine had a humble beginning as a large, clumsy, and (in terms of energy consumption) very inefficient device for pumping water out of deep coal and iron mines—something for which America had little need at that time. There are many claims about who first created a steam engine. The one invented by the English engineer Thomas Newcomen (circa 1710) and refined by James Watt (1775) had, by the 1780s evolved into a powerful, if inefficient, machine. The first engines produced only the simple linear motion of a piston in a cylinder; with the appropriate valves, they generated a linear pumping action. James Watt's innovations with respect to the condensing phase of the cycle within the cylinder greatly improved efficiency, but the real breakthrough came with adding a flywheel and an eccentrically placed connecting rod. That adapted the steam engine to produce rotary motion, which could then be applied for industrial use, first in steam-powered grist mills, saw mills, and stamping machines. These uses of steam were revolutionary in every sense. Soon there were plans for its application to ships and even carriages for the road.

Jefferson had the good fortune to be in Europe when some of the first changes in steam engines were made. For example, in Paris, steam-powered pumps were used to propel water from fire engines to put out blazes. Two single-acting Watt engines were also used to pump water to reservoirs that fed houses on the Right Bank, including Jefferson's Hôtel de Langeac, which was very modern, with a ground-floor bathroom and water closets on the upper floors.

During a trip to London in 1786, Jefferson visited Matthew Boulton's New Albion Mills, near Blackfriars Bridge, which used two Boulton-Watt steam engines to drive eight pairs of stones to grind 150 bushels of flour an hour. This was vastly faster than a water mill, which could grind only some 10 pounds per hour.[15] Jefferson wrote to Charles Thomson: "I could write you volumes on the improvements which I find made and making here in the arts. One deserves particular notice, because it is simple, great, and likely to have extensive consequences. It is the application of steam, as an agent for working grist mills. I have visited the one lately made here. It was at that time turning eight pair of stones. It consumes 100. bushels of coal a day. It is proposed to put up 30. pair of stones. I do not know whether the quantity of fuel is to be increased. I hear you are applying this same agent in America to navigate boats, and I have little doubt but that it will be applied generally to machines, so as to supersede the use of water ponds, and of course to lay open all streams for navigation. We know that steam is one of the most powerful engines we can employ; and in America fuel is abundant."[16]

Thomson wrote back asking whether the steam was used directly to power the mill or whether it was used to lift up water, which then drove the machinery. It was the obvious question, for steam had so far principally been used for pumping. Jefferson replied that steam was the "immediate mover" and was now "beginning to be applied to the various purposes of which it is susceptible."[17] In a postscript to this letter, Jefferson devoted his usual attention to the arithmetic details. Having talked with Boulton, he wrote: "He compares the effect of steam with that of horses in the following manner. 6 horses, aided with the most advantageous combination of the mechanical powers hitherto tried, will grind six bushels of flour in an hour; at the end of which time they are all in a foam, and must rest. They can work thus 6. hours in the 24, grinding 36 bushels of flour, which is 6. to each horses for the 24 hours. His steam mill in London consumes 120 bushels of coal in 24 hours, turns 10 pr. Of stones, which grind 8 bushels of flour an hour each, which is 1920 bushels for the 24 hours. This makes a peck and a half of coal perform exactly as much as a horse in one day can perform."

As for steam navigation, by 1783 the steam-powered boat was already a reality; the Marquis de Jouffroy had run one (just barely) at Lyon. It was powered by a Newcomen-model engine but essentially shook itself apart. Because Jefferson was in Paris, he missed the Constitutional Convention in Philadelphia in 1787. For that event John Fitch demonstrated on the Delaware River a forty-five-foot steam-paddle vessel that he had built. Two years later, Jefferson reported to Thomas Paine, "There is an Abbé Arnal at Mismes who has obtained an exclusive privilege for navigating the rivers of this country by the aid of the steam engine."[18]

After his return to America, Jefferson's interest in steam continued, and in 1799, Robert Livingston sent him a design for an improved steam engine.[19] When Jefferson was appointed secretary of state, he chaired the new Patent Board for three years. Although he had once thought that the idea of patents was bad because it would allow monopolies on ideas and put a brake on innovation, at the Patent Board he saw the immense value of the patent system in encouraging the development of new ideas. Among the proposals the board reviewed in Jefferson's time were several for the development of steamboats—notably the rival claims of John Fitch and James Rumsey.

While in Paris, Jefferson came across another intriguing, if not faintly comical, invention: the application of a very large, hand-powered screw propeller to a small boat. It was not the modern kind of propeller but a helical contraption for rowing through the air. Jefferson examined the device minutely, in his typical way, and described it for several of his American correspondents. The screw was "about 8. feet long, it's axis is of 9. inches diameter and the spiral vane of about two feet radius. very large . . . the thread of which was a thin plate two feet broad applied by it's edge spirally around a small axis . . . This, turned on its axis in the air, carried the vessel across the Seine."[20] To Hugh Williamson he noted: "Being fixed on a couple of light boats and turned rapidly on it's axis (by means of a wheel, band, and pulley) it carried them across the Seine very quickly."[21]

The scientific Jefferson immediately saw that such a device would work very much better in water than in air: "it's effort [is] proportioned to the want of tenaciousness or resistance in the fluid," but "it might be useful for moving a ballon." The Reverend James Madison wrote to Jefferson agreeing that the helical screw might be applied to balloons, but Jefferson had already moved on. The helical propeller had reminded him of the (failed) invention of an underwater vessel—a submarine-torpedo, that David Bushnell tried to use against British ships in the Revolutionary War. He wrote off to George Washington to be reminded of the details of the torpedo; Washington could not remember much of the case.[22]

A minor invention that took some of Jefferson's time was a kind of oil lamp said to have been invented by a M. Aimé Argand, "in which the flame is spread into a hollow cylinder, and thus brought into contact with the air within as well as without. Doctr. Franklin had been on the point of the same discovery. The idea had occurred to him; but he had tried a bull-rush as a wick, which did not succeed. His occupations did not permit him to repeat and extend his trials to the introduction of a larger column of air than could pass through the stem of a bull-rush."[23]

Jefferson bought copies of the lamp in London to send one to the Reverend Madison and one to Benjamin Vaughan for the American Philosophical Society. The lamp gave much light for little oil. "There is but one critical circumstance in the management of it; that is the length of the wick above the top of the cylinder. If raised too high it fills the room with smoke. If not high enough it will not yeild it's due light. The true medium is where it first ceases to give a sensible smoke in the room. Two or three experiments will set you to rights in this. I send some spare wicks, and a set of spare glasses."[24]

Jefferson was at that time very much opposed in principle to the patent system and was happy to learn that others than M. Argand (in both London and Philadelphia) were selling versions of the lamp. He was disappointed, however, when he sent back home details of the new "phosphoretic matches" (predecessors of modern matches) only for Charles Thomson to tell him: "The phosphoretic matches I have seen. They are sold in our toy shops."[25]

Among other minor issues, he reported to Rittenhouse that the "Abbé Rochon has lately made a very curious discovery in optics. He has made lenses with a chrystal from Iceland which have a double focus, perfectly distinct and at considerable distances from each other. I looked thro' a telescope to which he had adapted one of these and saw at the same instant an object on the banks of the Seine and a house half a mile further back with equal precision, the intermediate objects being dim as not at a proper distance for either focus." Abbé Rochon had also "lately made a telescope with the metal called Platina, which while it is susceptible of as perfect a polish as the metal heretofore used for the specula of telescopes, is inattackable by rust as gold and silver are."

Also, "A windmill has lately been announced which unites these advantages. 1. It offers itself to the wind in whatever direction that is. 2. It admits an easy change in the inclination of it's wings, and may therefore be readily adapted to it's greater or lesser force. 3. It substitutes wood instead of canvas for the vanes. 4. The geer is all fixed, and not moving on a center as is usual. These are the only improvements in the arts which I recollect at present as worth mention." But then he added: "There is a person here who has hit on a *new method of engraving*. He gives you an ink of his composition. Write on copper plates any thing which you would wish to take several copies; and in an hour the plate will be ready to strike them off. So of plans, engravings &c. This art will be amusing to individuals if he should make it known."[26]

A single exchange of letters between Jefferson and Madison in late 1785 and early 1786 gives a sample of the remarkable range of other subjects that Jefferson explored. In addition to steam power and the helical screw, he described double stars and other astronomical observations, the exciting report of a supposed new "Planet Herschel" (Uranus), the question of the density of the atmosphere, the propagation of sound, methods of making hydrogen gas from pit coal for use in balloons, and machines for copying letters.[27]

Jefferson wrote: "In the last volume of the Connoissance des tems you will find the tables for the planet Herschel. It is a curious circumstance that this planet was seen 30 years ago by Mayer, and supposed by him to be a fixed star. He accordingly determined a place for it in his catalogue of the Zodiacal stars, making it the 964th. of that catalogue. Bode of Berlin observed in 1781 that this star was missing. Subsequent calculations of the motion of the planet Herschel shew that it must have been, at the time of Mayer's observation, where he had placed his 964th. star." (*Connaissance des Temps* was, and still is, an French official newsletter on astronomy.)

Jefferson next cited the *Transactions* of the American Philosophical Society. "Herschell has pushed his discoveries of double stars now to upwards of 900. being twice the number of those communicated in the Philosophical transactions. You have probably seen that a Mr. Pigott had discovered periodical variations of light in the star Algol. He has observed the same in the η of Antinous, and makes the period of variation 7D. 4H. 30′, the duration of the 576 increase 63.H. and of the decrease 36H. What are we to conclude from this? That there are suns which have their orbits of revolution too? But this would suppose a wonderful harmony in their planets, and present a new scene where the attracting powers should be without and not within the orbit. The motion of our sun would be a miniature of this. But this must be left to you Astronomers."

Madison's reply shows how detailed and technical their discussions about their shared interest, astronomy, became. "The Circumstance you mentioned relative to the Planet Herschel is indeed a curious one . . . I had not before heard of the Observations of Mr. Pigott. They seem to be more precise than any before made upon that remarkable Phænomenon of a periodical Variation of Light in some of the Stars. As early as 1596, one in the Neck of the Whale was observed to increase and decrease regularly, it's Period being about 334 Days. But I have seen no Account of any Stars whose periods were so short, or so well ascertained as that you mention. Such Phænomena undoubtedly afford ample Room for Philosophical Speculation. Perhaps, like our Sun, those stars have a Rotation upon their Axes, and also their Maculæ, the larger and more constant, which, during their rotatory Motion, may cause a regular variation in their Light, in proportion as those Maculæ are more or less turned to the Observer. Or, perhaps as Maupertuis supposes, that Appearance may arise from their Form. They may be considerably flattened, so that they will appear more or less luminous, as the broad or narrow Side is turned towards us. Your Idea that they may be Suns, which have their orbits of Revolution is more extensive than either of those Suppositions; and it is not the Part of a Philosopher to deny the Possibility of certain Dispositions in Nature, when the Phænomena seem to indicate them, merely because he cannot fully comprehend the manner in which they act. It seems to be doubtful whether Men are more slow in collecting the Laws of Nature, or in applying them when collected and known."

As for the Icelandic crystal, Madison apologized. "I have been since sorry that I proposed those Queries respecting the Chrystal. It was giving you a Trouble which I ought to have spared. I have lately seen some Account of the Icelandic Chrystal, as long known to Opticians for its singular refractive Powers. It is described as a Kind of Talc, found in the form of an oblique Parallelopiped, and composed of Lamina which will cleave parallel to either of it's Sides; from which Constitution, its Property may probably be derived. Dr. Franklin brought a Peice of this Chrystal to Phila. which he gave to a Friend of mine who informed me of it's Property of giving a double Image. They have also in Phila. one of the Telescopes made of it, by which the Suns apparent Diameter is readily measured, and small Distances accurately determined."

A subject of interest in Parisian scientific circles was the work of a Doctor Ingenhousz, who claimed that growth in plants could be stimulated by electric currents. Jefferson bought a copy of Ingenhousz's pamphlet on the subject but later reported to the Reverend James Madison that "he now however retracts it, and finds, by more decisive experiments, that the electrical fluid can neither forward nor retard vegetation. Uncorrected still of the rage of drawing general conclusions from partial and equivocal observations, he hazards the opinion that *light* promotes vegetation."[28]

A more controversial subject was the promotion by Franz Anton Mesmer (1734–1815), a German physician, of what he called animal magnetism, which was connected with the flow in the body of an undefined cosmic fluid, and his claims for the value of hypnosis in medicine. Disease, he said, was due to an obstruction in the flow of the magnetic "fluidum," and he claimed to be able to restore that free flow within the body and between the body and the cosmos. Mesmer created a sensation by putting patients into a trance and effecting "cures." Just as Jefferson arrived in Paris, Louis XVI set up a commission to investigate the claims. The commission consisted of the chemist Antoine Lavoisier, Dr. Joseph-Ignace Guillotin, the astronomer Jean Sylvain Bailly, and Benjamin Franklin. They could not find any evidence of a special fluid; the cures that Mesmer claimed were due to imagination. Only later was hypnosis (often called mesmerism) developed more fully as a diagnostic and therapeutic tool. Mesmer himself was pronounced a charlatan.

Fraud or not, Mesmer's influence was already being felt in America. The Marquis de Lafayette was persuaded that his cures were genuine and made a presentation to the American Philosophical Society on the subject.[29] When Hugh Williamson wrote asking for details of this apparently exciting development, Jefferson stuck a pin in the balloon. "The doctrine of animal magnetism after which you enquire is pretty well laid to rest. Reasonable men, if they ever paid any attention to such a hocus pocus theory, were thoroughly satisfied by the Report of the commissioners. But as the unreasonable is the largest part of mankind, and these were the most likely to be affected by this madness, ridicule has been let loose for their cure: Mesmer and Deslon have been introduced on the stage, and the contest now is who can best prove that they never were of their school."[30]

Jefferson sent a copy of the commission's report to Charles Thomson, who replied: "The report you sent me has removed this doubt and though it has sufficiently demonstrated that Mr. Mesmer and his disciples have discovered no new property in nature yet it has itself made a very wonderful and very important discovery, namely to what degree the imagination can operate on the human frame."[31]

Of all the sciences, Jefferson had a less wholehearted, more mixed appreciation of chemistry, the more modern versions of which were still in their infancy. "I do not know whether you are fond of chemical reading," he wrote to Rittenhouse. "There are some things in this science worth reading."[32] It was just then that attempts were being made to develop a rational system of chemical classification and nomenclature. "A schism . . . has taken place among the chimists. A particular set of them here have undertaken to remodel all the terms of the science, and to give to every substance a new name the composition, and especially the termination, of which shall define the relation in which it stands to other substances of the same family. But the science seems too much in it's infancy as yet for this reformation."[33]

A common contemporary attitude toward chemistry was expressed by Buffon. Jefferson reported, "Speaking one day with Monsieur de Buffon on the present ardor of chemical enquiry, he affected to consider chemistry but as cookery, and to place the toils of the laboratory on a footing with those of the kitchen." But Jefferson disagreed. "I think it on the contrary among the most useful of sciences, and big with future discoveries for the utility and safety of the human race. It is yet indeed a mere embryon. It's principles are contested. Experiments seem contradictory: their subjects are so minute as to escape our senses; and their result too fallacious to satisfy the mind. It is probably an age too soon to propose the establishment of system."[34]

A quarter of a century after that 1788 conversation, matters had not changed very much, and Jefferson echoed Buffon's thoughts. "I have wished to see chemistry applied to domestic subjects, to malting, for instance, brewing, making cider, to fermentation and distillation generally, to the making of bread, butter, cheese, soap, to the incubation of eggs, &c."[35] In fact, major advances in chemistry, and particularly in biochemistry, did indeed flow from such investigations, exemplifying Jefferson's belief that theoretical and practical science went hand in hand. However, chemistry was not a subject he valued highly as a pursuit for himself or his grandson. "Chemistry . . . is the least useful & the least amusing to a country gentleman of all the ordinary branches of science . . . for chemistry you must shut yourself up in your laboratory . . . chemistry is of value to the amateur inhabiting a city."[36]

Fascinating as Paris was, especially to someone like Jefferson, who was always at least a closet revolutionary, it was not enough to contain Jefferson's restless mind. In 1786 he and John Adams made a seven-week journey around Britain, looking, among other things, at the gardens made popular in a book by John Whately.[37] He and Adams visited them all, along with Oxford University and of course the London shops.

In 1787, Jefferson set off on a three-month journey through southern France and Italy.[38] Here he was transported by the pleasure of seeing for the first time the architecture of the Romans. He wrote to Madame de Tessé, indulging his fancies: "Were I to attempt to give you news, I should tell you stories a thousand years old. I should detail to you the intrigues of the courts of the Caesars, how they affect us here, the oppressions of their Praetors, Praefects &c. I am immersed in antiquities from morning to night. For me the city of Rome is actually existing in all the splendor of it's empire. I am filled with alarms for the event of the irruptions dayly making on us by the Goths, Ostrogoths, Visigoths and Vandals, lest they should reconquer us to our original barbarism." Typically the letter ends with a little gallantry: "If I am sometimes induced to look forward to the eighteenth century, it is only when recalled to it by the recollection of your goodness and friendship."[39] But his purpose in this journey was only part touristic; part was to be a sort of economic spy.

Jefferson reported back to the United States on a huge range of agricultural topics, from the cultivation of crops and their prices, viticulture and winemaking, the design of plows, manuring and the care of fields and soils, steam-powered grist mills, methods of cheese making, and the design and operation of France's extensive and ambitious canal system. He sought to discover the secret of the superiority of Italian rice over that from the Carolinas. "I wished particularly to know whether it was the use of a different machine for cleaning which brought European rice to market less broken than ours."[40] Since the machines were identical to those used in America, the difference had to lie in the cultivars of rice, which were jealously guarded. "It is a difference in the species of grain, of which the government of Turin is so sensible, that, as I was informed, they prohibit the exportation of rough rice on pain of death . . . I have taken measures however for obtaining a quantity of it which I think will not fail & I bought on the spot a small parcel which I have with me." And eventually the growing of upland rice, rather than the lowland variety, which required swamps (and their concomitant insects and diseases and hideous conditions for slave workers) was experimented with in America. Jefferson even grew upland rice at Monticello.

Jefferson also took time to explore an interesting report that was making the rounds. A certain M. de la Sauvagere claimed to have observed petrified shells growing within rock without any connection to living animals. Sauvagere was vouched for by many, including Voltaire, and Jefferson was circumspect in his judgments after having spoken to people in and around Tours, where the growth was supposed to have been seen. It did not to seem to him to be against any law of nature, but nonetheless it seemed improbable. He could not make himself either believe it or dismiss it. The difficulty was that there were no conclusive rival explanations of the fact that perfectly real-looking seashells could be found embedded in mountain rocks. "While I withhold my assent, however, from this hypothesis, I must deny it to every other I have ever seen, by which their authors pretend to account for the origin of shells in high places."[41]

In 1788, Jefferson followed his southern tour with an eight-week tour through Holland and the Rhine Valley, where he concentrated on observing the wine industry and the engineering of canals.[42]

Canals, and efficient designs for canal locks, were an important factor in transport in Europe and promoted the growth of industrialization. Two other inventions, above all, made possible the Industrial Revolution. Steam power was the first; it both allowed recovery of coal from deep mines that would, without pumping, have flooded, and served as a source of power for machines. The second innovation was mass production using steam engines, and Jefferson, in Paris, was able to observe and report on this also.

In the 1770s the French gunsmith Honoré le Blanc established (perhaps for the very first time) the basic principles of mass production, first in the manufacture of musket locks—the device of trigger and flint that caused the powder to discharge. He established methods of making the separate parts using jigs, molds, and dies, all the pieces being effectively identical (or at least within an acceptable range of variation) and thus interchangeable. Le Blanc had invented the principle of the *spare part.* Previously, because every gun was custom-built, any broken part had to be laboriously and expensively made by a craftsman. Now a part that would fit could simply be taken from a bin.

While the advantage of le Blanc's system was that a gun lock could easily be built or repaired, the disadvantage was that it would put skilled craftsmen—except the die and jig makers—out of work. So le Blanc had difficulty in persuading French industry to adopt his ideas. Jefferson, however, saw all the advantages of such a system in America, where labor was often scarce, and wrote to John Jay: "An improvement is made here in the construction of the musket which it may be interesting to Congress to know": it "consists in the making of every part of them so exactly alike that what belongs to any one, may be used for every other musket . . . I went to the workman, he presented me the parts of 50. locks taken to pieces . . . I put several together myself taking peices at hazard as they came to hand, and they fitted in the most perfect manner."[43]

Jefferson later wrote about the principle of mass manufacture to Henry Knox, then secretary of war, reporting that le Blanc had proceeded from the manufacture of locks alone to making entire muskets according to these principles. So that Knox could try them for himself, he sent the parts for "half a dozen of his officers fusils . . . You will find with them, tools for putting them together, also a single specimen of his souldier's lock. He formerly tolld me, and still tells me, that he shall be able, after a while, to furnish them cheaper than the common musket of the same quality."[44]

Traditional craftsmen opposed mass manufacture, and European countries were slow to take up the idea. Le Blanc told Jefferson, "If the situation of the finances of this country should oblige the government to abandon him, he would prefer removing with all his people and implements to America." Jefferson clearly would have liked this to happen, but le Blanc died soon thereafter. The idea of mass production soon took off in America, especially when, in 1798, the inventor of the cotton gin, Eli Whitney, obtained a $134,000 contract from Congress for the mass manufacture of twelve thousand muskets.

American and Yankee ingenuity turned le Blanc's principle into a force for the making of all manner of devices. Whitney's factory had machines for boring barrels and shaping wooden rifle stocks, as well as for making the parts for the locks. Once precisely engineered parts could be produced by the appropriate machine tools, it was possible to produce en masse a range of devices that had previously depended on the skill of individual craftsmen—cheap, reliable clocks and watches being a particular case in point. Eli Terry started making manufactured wooden clocks in Waterbury, Connecticut; his assistant Seth Thomas continued the work, and then Chauncey Jerome, using brass, made as many as five hundred movements a day. Soon America was supplying clocks all over the world.

It is perfectly in accord with Jefferson's interest in the practical and technical arts that he should have established at Monticello a profitable nail manufactory. It was another example of mass production—in this case, nails from 6-penny to 20-penny sizes, tacks, and brads were made from iron wire and bars. The nailery was not highly mechanized; its mass production depended on repetitive hand labor at two forges. At its peak, though, a dozen or so teenage slave boys produced between five thousand and ten thousand nails per day. (Although a nail-cutting machine was bought later for use on the smaller sizes, it gave a lot of trouble.) The nailery was one of Jefferson's better commercial ventures; he earned around two thousand dollars per year from the sales of nails around the region (his customers included James Madison, who needed nails for the building of Montpelier). It is ironic that Jefferson, who for so long stood for the value of the agrarian life and the virtue of the individual farmer and craftsman, should have played even a small part in an American Industrial Revolution that ran counter to so many of those ideals.

CHAPTER THIRTEEN

The Practical Scientist

ONCE BACK HOME IN VIRGINIA, Jefferson was able to turn the inventiveness of European science to his own use. Today, when the words "Jefferson" and "science" are brought together, the response of most people is to conjure up the man who was famous for his inspired attention to every kind of mechanical device and technological innovation. That is the Jefferson familiar to the general public; his mechanical genius is demonstrated, with justifiable pride, in tours of his magnificent home, Monticello. Unfortunately, it has to be recorded that the number of devices that Jefferson actually *invented* is very small; he borrowed and adapted many more. But that does not matter; there was much more to Jefferson's philosophy than tinkering. What he was able to do was to take existing ideas and devices and improve them, bringing to bear his depth of learning, his wide travels, and, in some cases at least, his ability in mathematics.[1] Just to have improved one or two objects would have been enough for ordinary mortals to procure fame, if not fortune.

In his later years (especially during and after the presidential years, 1801–1809), Jefferson's scientific interests shifted. Books on science had been the largest single category in his great library but when it was sold and he assembled his "Retirement Library," scientific books now accounted for only 9 percent of the total 931 titles (as listed in the catalogue of its sale in 1829).[2] The library of some 350 titles that he created for his house at Poplar Forest, another retirement project, had only 8 titles in natural science (but that included the full 52 volumes of Buffon's *Histoire Naturelle*).

By this stage he had clearly moved away from some of his more theoretical interests in natural philosophy, considering contemporary speculation on the origins and history of the earth, for example, as "too idle to be worth a single hour of any man's life."[3] At the same time, he became more and more preoccupied with mechanical devices, particularly those for the improvement of Monticello and his farms. Later he poured his architectural skill into designing the new University of Virginia. He not only helped found the university but designed its buildings and advised on its curriculum.

It is remarkable that he was involved with so many technical issues and so many kinds of mechanical devices and technological innovations. One does not have to go any further for a demonstration of the extraordinary range of Jefferson's technical turn of mind than the entrance hall of Monticello. After resigning as secretary of state in 1793, Jefferson started to redesign the house. Tearing down of the "old" building began in 1796, just before he was elected vice president. Jefferson put into the design of the new Monticello, with its wonderful dome and extensive "dependencies" (outbuildings and services), his refined ideas on architecture. Inside, he gave full rein to his inventiveness, now revealed after painstaking work by the curatorial staff at the house.

One enters the house from a broad porch, over which hangs his Great Clock. By the late eighteenth century, the clock—a crucial element of domestic economy—was finally being made to work more or less reliably. Jefferson gave specifications to a clockmaker named Peter Struck in 1792 for a large clock to be installed at Monticello. It has two faces, one outward to the porch and the other inward to the entrance hall. The outer face shows only the hours, and there was a loud bell that sounded hourly to let the fieldworkers know the time. The inner face shows the inhabitants of the house both hours and minutes. It is powered by falling weights, essentially six small cannon balls on a rope that passes over pulleys to drop down in the corner of the hall. Jefferson put signs on the adjacent wall so that, as the weights dropped, they indicated the days of the week. This clock was the kind usually installed in a tower; its weights needed to move up and down a long distance. So Jefferson cut a hole in the hall floor so that every week the cannon balls could disappear belowstairs during late Friday and Saturday. Jefferson also devised a folding ladder to reach the clock's works to wind it once a week.[4]

In this view of the front hall at Monticello today, the weights for the big clock and the plaques for the days of week can be seen in the corner, together with part of Jefferson's reconstructed "museum" collections. Courtesy of the Thomas Jefferson Foundation, Inc., at Monticello. Photograph by Charles Shoffner.

The Entrance Hall was Jefferson's personal scientific museum—filling up over the years with Indian artifacts and archeological objects, maps, his mastodon fossils, and a collection of animal heads. From it, the visitor moves to the Parlor through a pair of doors, linked mechanically by a chain mechanism to open as one. The Dining Room is separated from the Tea Room by another double-door arrangement, this a pair of sliding pocket doors. Both kinds of doors were probably copied by Jefferson from versions seen in Paris.

Jefferson's study holds his famous four-sided revolving book stand. A cube with four inclined rests for books on its sides and another on top, it allowed Jefferson to consult five works at the same time. The book stand might truly be something that Jefferson designed rather than copied. It was

Jefferson's revolving book stand was one of the devices he really did invent. Because it essentially consists of five music stands on a revolving base, it allowed him to consult five books at once. Courtesy of the Thomas Jefferson Foundation, Inc., at Monticello. Photograph by Edward Owen.

made on site in 1810. Jefferson evidently liked revolving objects. In Philadelphia in 1776 he bought a swivel chair to which he later, with typical inventiveness, added a writing pad.

Between the Dining Room and the servants' hall was a lazy Susan that allowed dishes to be passed to and fro without servants needing to enter the room. There was also a small elevator waiter next to the fireplace that transported wine bottles from the cellar. We would call the latter a dumbwaiter, but in Jefferson's time that term was used for a wheeled side table with shelves. For small dinner parties, Jefferson would place one such dumbwaiter by each guest, and on it would be all the dishes for the meal. The guest would place the used dishes back on the shelves. This in France was called an étagère. Another revolving device that he liked was a moveable hanging device for inside a clothes closet. If he did not invent all these devices, he made the drawings for hired joiners like Thomas Walker and James Dinsmore and the skilled slave John Hemings to construct them in the Monticello workshops.

In his Cabinet (study) was Jefferson's polygraph machine, which he used to make a copy of each of his letters as he wrote it. This machine was invented by an Englishman, John Isaac Hawkins, and was improved by Jefferson's friend Charles Willson Peale. The polygraph held two pens in a pantograph arrangement. As Jefferson wrote with one pen, the other moved automatically over a parallel sheet of paper. Fortunately for historians, the polygraph meant that copies of Jefferson's correspondence were kept together while the originals were dispatched to the four corners of the earth.

Jefferson acquired his polygraph in 1804; previously he had used a copying press. With a copying press, a sheet of dampened tissue paper could be pressed against a letter once it had been written. Several copies of a single letter could be made in this way. James Watt (deservedly more famous for his steam engines) patented such a copying press in 1780. The original machine, using either a roller or a vertical screw, was large and cumbersome. Jefferson designed a small portable version and had one made in London in 1787 along with a portable desk. Other versions were made for him at Monticello. (Possibly other inventors had the same idea for making a portable press, but Jefferson seems to have come to it independently.)

Contrary to popular lore, Jefferson did not invent macaroni and cheese or French fries ; he imported both ideas from France, although he might have been the first, or one of the first, to do so. Nor did he invent a machine to make spaghetti ("maccaroni"), although he made a careful study of one that he saw in Italy in 1787 and sent back a mechanical drawing of it and instructions for its use.[5]

Jefferson did not invent the pedometer, either, although he owned at least one and may have been the first to introduce pedometers to America. He wrote to James Madison in 1786 about this new device. "I remember your purchase of a watch in Philadelphia. If she should not have proved good, you can probably sell her. In that case I can get for you here, one made as perfect as human art can make it for about 24. louis . . . For 12. louis more you can have in the same cover, but on the backside, and absolutely unconnected with the movements of the watch, a pedometer which shall render you an exact account of the distances you walk."[6] The pedometer worked as a step-counting device, recording the number of times a small ball moved inside it. Calibrated to the length of the stride, the count would yield the distance traveled; the same principle is used in modern digital devices.

As someone devoted to keeping records of the climate wherever he was, Jefferson was particularly interested in the tricky question of how to make a reliable hygrometer for measuring humidity in the air. His thoughts on this subject show us his mechanical aptitude and ingenuity. At the time there were a number of different technical devices in use, all depending on the change in shape of strips of wood, ivory, or whalebone in response to moisture in the air. Jefferson obtained several different kinds of hygrometer and, with experimentation, discovered that there was not only a problem in the nonlinear response of strip of material but the absence of a definable standard against which devices could be compared.

As Jefferson wrote to Benjamin Vaughan in England, "I have been a little at a loss with the Hygrometer of De Luc you were so kind as to send me. It is graduated from zero to 100, and I had understood these were his extremes. Those of De Saussure are the same. Yet, while this of De Luc, exposed to the open air has never fallen below 26. nor risen above 55. since it was in my possession, those of De Saussure have been generally, during the wet spell we have had, at about 90."[7] His further correspondence on this issue shows how inventive his mind was.

Benjamin Franklin had in 1759 had the idea for a hygrometer using thin strips of wood after noting that the sliding lid of the box for a set of magnets that he had purchased sometimes moved smoothly and at others became stuck. Franklin saw the association of humidity with the lid's being stuck or not, and he quickly translated this observation into a suggestion for a measuring device. Essentially, a very thin strip of mahogany would be placed so that when it curled or straightened it would indicate "in sensible degrees, by a moveable hand upon a marked scale," changes in humidity.[8]

Both Jefferson and David Rittenhouse suggested improvements in Franklin's device, which was "liable to the objection . . . that equal extensions of the wood are not equally indicated on the dial plate. The fillet of wood expanding in the direction of the tangent of the circle, the circle should be divided into portions corresponding to equal parts of the tangent, that is, the index should point out, not the angle it has moved over, but it's tangent." Jefferson sketched a new system in which the dual wooden strip consisted of one with the grain lengthwise and the other crosswise, but he knew that the instrument was "not fit for comparative observations, because no two peices of wood being of the same texture exactly, no two will yeild exactly alike to the same agent."[9]

Jefferson improved other devices besides the hygrometer. Several of his improvements were associated with his farms. Hemp was grown at Monticello, as on many plantations, for cordage and for cloth for the slaves—a rough and scratchy cloth at that, even when blended with cotton. Jefferson saw that a device could be added to his sawmill (powered by the Rivanna River) to serve as a "hemp break." "I make the same saw-gate . . . move another lever at the other end of which is suspended the upper head-block of a common hemp break (but much heavier than common) . . . While two persons feed the break with the hemp stalks, a third holds the hemp already beaten and formed into a twist, under the head block, which beats it most perfectly . . . To make this work true, a section of a circle (like the felloe of a cart wheel, but shorter) is mortised on the end of the lever, with a groove in it for the suspending chain to lie in . . . I wish to make the same agent work an apparatus for fulling our homespun, but have not yet attempted it, tho' we need it much, as we clothe ourselves chiefly, and our laborers entirely in what we spin and weave in our family."[10]

One invention for which Jefferson can certainly claim full credit was his design for the moldboard of a plow—the part that lifts up and turns over the soil or sod. After observing European wooden plows, Jefferson wanted to improve on contemporary designs by making one that turned the sod over

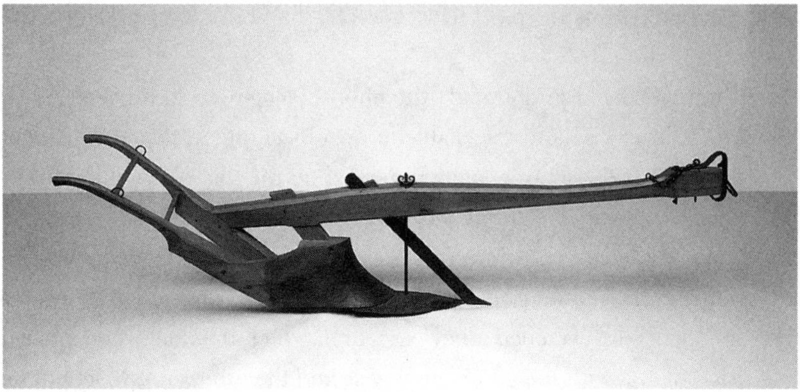

This model of Jefferson's design for an improved plow shows the shape of the wooden moldboard that was precisely calculated best to turn and lift the sod that the blade in front had cut. Courtesy of the Thomas Jefferson Foundation, Inc. at Monticello. Photograph by Edward Owen.

more completely and with less resistance. "It is on the principle of two wedges combined at right angles, the first in direct line of the furrow to raise the turf gradually, the other across the furrow to turn it over gradually. For both these purposes the wedge is the instrument of the least resistance."[11] Jefferson used his knowledge of calculus to combine the two wedges into one complex curve. "I have imagined and executed a mould-board which may be mathematically demonstrated to be perfect, as far as perfection depends on mathematical principles, and one great circumstance in its favor is that it may be made by the most bungling carpenter, and cannot possibly vary a hair's breadth in its form, but by gross negligence."[12]

The last claim might be overstated, but Jefferson had worked out not only the ideal form of the moldboard but the way any competent woodworker could saw one out of a block of wood. Later he had the moldboard cast in iron. Jefferson described his invention in detail in the *Transactions* of the American Philosophical Society and gained some acclaim, being awarded a gold medal by the French Society of Agriculture. But its use never really caught on, losing out to the conservatism of most farmers and being overtaken by further developments of the same sort.

Another device that was genuinely Jefferson's, even though others might have had the idea independently, was the spherical sundial. In 1809, Benjamin Latrobe had sent him a copy of the capital of one of the columns he had designed for the new United States Capitol. Jefferson put it on a suitable pedestal but thought it looked "bald for want of something to crown it." So he added a ten-and-a-half-inch globe of locust wood, but even then he "was not satisfied; because it presented no idea of utility. It occurred then that this globe might be made to perform the functions of a dial." He inscribed the suitable meridian lines, added a curved vane, and oriented it correctly. The result "enables me to judge within one or two minutes of solar time."[13]

One final device may be Jefferson's or at least was marked by Jefferson's inventiveness. We might expect cryptography to have appealed to his mathematical instincts. In times when all information passed by letter, to send anything reliably to Europe usually meant sending two or three copies by different ships. And still copies were lost to shipwrecks and pirates. Those that made it through were read by spies before they reached their destination; European postal officials routinely opened and read even the diplomatic mail. The long-standing solution was the use of code, and Jefferson sent many letters using simple codes during his Paris years and when he was secretary of state.

Mechanical code devices, rather like a kind of combination lock, were known in the fifteenth century. They may have been known to Jefferson, but his wheel cipher machine, to use the name it has acquired, though based on the same general principle, was considerably more complicated.[14] It consisted of a series of discs mounted on a spindle. Each had inscribed on its rim the letters of the alphabet in different, randomly scrambled order. The order in which the discs were lined up on the spindle was the first matter of importance. The machine was easy to use. The sender arranged the discs according to a previously defined order—say, by date. The discs were revolved until a single horizontal row of letters spelled out the required message. Then the sender picked out one of the other rows of letters, which would be a nonsensical series of letters. That nonsense

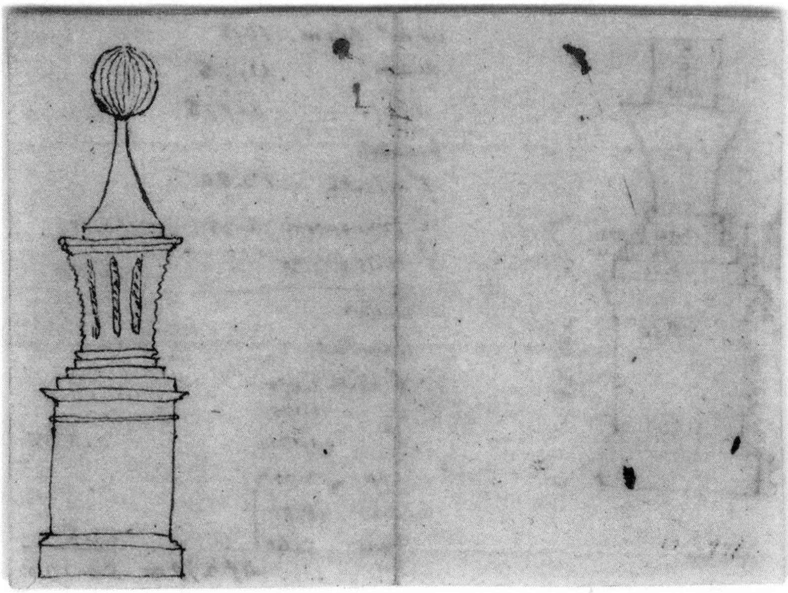

After Jefferson added a ball to improve an ornamental capital given to him by the architect Benjamin Latrobe, he realized that the ball could be used to create a spherical sundial. He sent this sketch in a letter to Latrobe. Monticello: pedestal. Drawing by Thomas Jefferson, [1816]. N147y; K149y. Original manuscript from the Coolidge Collection of Thomas Jefferson Manuscripts. Massachusetts Historical Society. Courtesy of the Massachusetts Historical Society.

stream became the message, which was sent by conventional methods. The recipient had to own an identical machine and had to know in advance the order of the discs for that date. Rotating the disks until the letters in one row spelled out the nonsense line made the correct message appear in another row. As with other cipher systems, maintaining the integrity of the system lay both in protecting the codebook in which the daily order of the disks was listed and receiving the nonsensical message accurately.

Jefferson may have had a wheel cipher constructed as early as 1793, when he was still secretary of state. His device was an invention far ahead of its time: it was another hundred years before the principle was rediscovered by Étienne Bazeries, a French military cryptanalyst. A version was in use by the United States Army in the early part of the Second World War,

The wheel cipher device was another of Jefferson's inventions; the whereabouts of his own machine is unknown. Courtesy of the National Cryptologic Museum.

and the same principles lay behind the famous German "Enigma" cipher machines. As with this device, much of what Jefferson devised or improved upon still has use today.

PART FIVE

The National Stage

CHAPTER FOURTEEN

Climate and Geography

ON JULY 1, 1776, THE thirty-three-year-old Jefferson had many weighty issues on his mind, but it was on that date, in Philadelphia, that he began to keep the diary of temperature and other weather details that he maintained on a daily basis for nearly fifty years. (The ever-competitive John Adams would no doubt have delighted in the longer record of a near-contemporary New Englander, Edmund Augustus Holyoke, of Salem, Massachusetts, who kept a similar record for seventy-five years.)[1] Among other things, Jefferson recorded that on July 4 the weather in Philadelphia was very pleasant, with a temperature of only 76 degrees Fahrenheit.

The reason for Jefferson's late start in making weather observations—a surprising delay in someone so devoted to measurements and lists of data—may have been that he did not own a thermometer until 1776. In colonial America, even such day-to-day commodities as glassware, silver, spices, and medicines had to be imported, usually from England. Most newspapers carried advertisements from shopkeepers who were selling a newly arrived cargo of domestic goods. Items like scientific instruments and particular book titles were rarely brought in that way; any unusual item had to be specially ordered. Everything took time. When the Reverend James Madison dispatched an order to London to replace the thermometer that British soldiers had stolen during the war—"It is impossible to procure one here"—he expected to wait two to three months for it to arrive.[2]

But sometimes, against the odds, a merchant would speculate by bringing in something unusual; so it must have been when Jefferson stopped in at John Sparhawk's London Bookstore on Market Street in Philadelphia. A thermometer was for sale, and he bought it for three pounds and fifteen shillings.[3] He had tried to obtain one much earlier. In 1769 he commissioned a fellow Virginian, James Ogilvie, who was traveling to Scotland to prepare for the church, to send one back, along with other items on a typically Jeffersonian wish list: two pairs of "rising dovetail door hinges," two Chinese brass bells "with wire and cranks," a case of bottles and canisters, strings for a violoncello, pullies and cord for Venetian blinds, locks, a "Scotch carpet," and a "Calendar" for pressing clothes.[4] But Ogilvie seems not to have been a reliable broker. When the Reverend Ogilvie returned to Virginia, he brought none of Jefferson's items with him.

An unknown number of Americans had owned thermometers before the Reverend Madison and Jefferson. Holyoke started his measurements in 1754. Benjamin Franklin gave Ezra Stiles at Yale College a thermometer in 1762. Most farming families kept records in their diaries; even anecdotal information was useful in determining planting and harvest times. George Washington kept a weather record for Mount Vernon for many years without using a thermometer. Whether Jefferson's longtime friend the Reverend James Madison had had a thermometer before 1770 is unknown, but someone in Williamsburg did, because Jefferson remarked in *Notes on the State of Virginia* that in 1766 the summer temperature had reached 98 degrees Fahrenheit. Probably the observer in this case was either Madison or their mutual friend Governor Fauquier.

Jefferson went beyond simple observation. He understood the special value in a consistent series of temperature recordings for probing into the underlying patterns of climate, which exemplified the difference between recording the weather as a hobby and recording it as a science. By linking his climate data to American geography in *Notes,* he can fairly claim to have been a pioneer in both fields. Although he wrote descriptively about geography in the early chapters of *Notes,* in the seventh chapter—"Climate"—he linked geography and meteorology analytically, which is to say, scientifically.[5]

Jefferson began his treatment of climate in *Notes* by reducing a run of five years (1772–1777) of daily measurements of temperature, rainfall, and wind direction taken at Williamsburg to monthly averages. The Williamsburg data were not Jefferson's, however. He was dividing his time between

Williamsburg and Monticello in those years and did not own a thermometer. They probably came from the Reverend Madison; the end of the run would seem to correspond with the loss of Madison's thermometer to British troops.[6] Jefferson did have data of his own when writing *Notes,* but they were less complete than those of Madison, who recorded wind direction and barometric pressure as well as temperature on a daily basis (when he had the instruments to do it). Jefferson, at that time, only recorded the temperature.

In 1784, at Jefferson's urging, Madison managed to borrow another thermometer and started keeping records again so that they could compare conditions in Williamsburg and Monticello.[7] This instrument, Madison reported, "appears very sensible as to Heat or Cold, tho' it is so constructed that I cannot ascertain the Accuracy of the Division by plunging it in boiling water . . . In order to observe the greatest Cold, I expose the Thermometer to the Night Air. It either stands out all Night, or lies in an open window. The remainder of the Day it is also as much exposed as it prudently can be, on Account of the Danger of meeting with some Accident—for it is borrowed. I Mention this, because, both ought to be treated in the same Manner. I wish we had [a] Barometer but there is no Possibility of getting one here at present."[8]

Jefferson refined his own record-keeping practices; in 1790 he described them in a letter from New York to his son-in-law Thomas Mann Randolph, typically trying to enlist him in the effort. "I will propose to you to keep a diary of the weather here and wherever you shall be, exchanging observations from time to time. I should like to compare the two climates by cotemporary observations. My method is to make two observations a day, the one as early as possible in the morning, the other from 3. to 4. aclock, because I have found 4. aclock the hottest and day light the coldest point of the 24. hours. I state them in an ivory pocket book in the following form, and copy them out once a week. The 1st. column is the day of the month. The 2d. the thermometer in the morning. The 4th do. [ditto] in the evening. The 3d. the weather in the morning. The 5th do. in the afternoon. The 6th is for miscellanies, such as the appearance of birds, leafing and flowering of trees, frosts remarkably late or early, Aurora borealis, &c. In the 3d. and 5th. columns, a. is *after:* c, cloudy: f, fair: h, hail: r, rain: s, snow. Thus c a r h s means, *cloudy after rain, hail and snow.* Whenever it has rained, hailed or snowed between two observations I note it thus, f a r (i.e. fair after rain) c a s (cloudy after snow &c.) otherwise the falling weather would escape notation. I distinguish weather into fair or cloudy, according as the sky is more or less than half covered with clouds. I observe these things to you, because in order that our observations may present a fair comparison of the two climates, they should be kept on the same plan. I have no barometer here, and was without one at Paris. Still if you chuse to take barometrical observations you can insert a 3d morning column and a 3d afternoon column."[9]

Wherever he found himself—Monticello, Annapolis, Philadelphia, New York, Paris, London—Jefferson continued to make daily records of temperature, and he instructed his daughters to continue the Monticello sequence when he was away. He also encouraged friends and acquaintances to keep their own records. He wrote to the future president James Madison, "I wish you had a thermometer. Mr. Madison of the college and myself are keeping observations for a comparison of climate. We observe at Sunrise and at 4. o'clock P.M. which are the coldest and warmest points of the day. If you could observe at the same time it would shew the difference between going North and Northwest on this continent. I suspect it to be colder in Orange or Albemarle than here."[10] When Madison did not reply, he wrote again.[11] Once Madison had obtained a thermometer, the two friends continued to experiment with the best places to site their instruments and the time of day to make their observations.[12]

He wrote similarly to Isaac Zane, a Virginia ironmaster, asking him to record temperatures in his icehouse and at his spring. He developed a correspondence with William Dunbar, a Scottish immigrant plantation owner in Florida, both discussing meteorological observations and asking for vocabularies of Indian languages.[13] Nor were his overseas friends spared. He wrote to the Italian naturalist and chemist Giovanni Fabbroni in 1778: "I have yet a little leisure to indulge my fondness for philosophical studies. I could wish to correspond with you on subjects of that kind. It might not be unacceptable to you to be informed for instance of the true

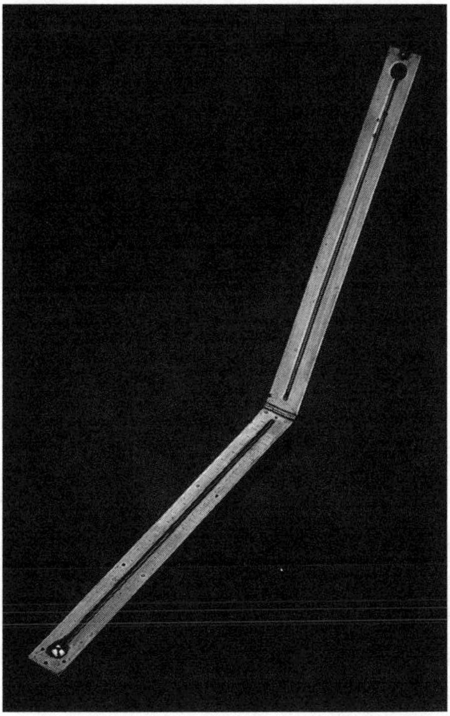

This traveling thermometer was owned by Jefferson; its case folded for protection. Courtesy of the Division of Political History, National Museum of American History, Smithsonian Institution.

power of our climate as discoverable from the Thermometer, from the force and direction of the winds, the quantity of rain, the plants which grow without shelter in the winter &c. On the other hand we should be much pleased with cotemporary observations on the same particulars in your country, which will give us a comparative view of the two climates."[14]

In his *Notes* chapter on climate, Jefferson also reported on wind direction, comparing conditions in the piedmont (Monticello) and coastal plain (Williamsburg) with averages from the two sites. However, this was not a true comparison. Madison was not able to provide a nine-month run for the same dates at Williamsburg as Jefferson had for Monticello (probably from 1783 to early 1784). So Jefferson used the averages from the 1772–1777 Williamsburg data. Even so, Jefferson's conclusion was accurate: while "the South-west wind prevails equally at both places" at the coast, the next most prevalent wind was from the northeast, while in the mountains it was from the northwest.

An additional difference between the winds at Monticello and Williamsburg was that those at the coast were more humid, "heavy and oppressive to the spirits," while the northwest wind was "dry, cooling, elastic and animating." This was just one instance of Jefferson's careful attention to the attractiveness of different climates. He never gave up claiming that Monticello had the most salubrious climate in the Unites States whereas that of the Carolinas was positively unhealthy. Among other differences between Williamsburg and Monticello was the incidence of frosts and dews, both being less marked at Monticello.

Jefferson's overall conclusion about the American climate was this: "In an extensive country, it will of course be expected that the climate is not the same in all its parts. It is remarkable that, proceeding on the same parallel of latitude westwardly, the climate becomes colder in like manner as when you proceed northwardly. This continues to be the case till you attain the summit of the Alleghaney, which is the highest land between the ocean and the Missisipi. From thence, descending in the same latitude to the Missisipi, the change reverses; and, if we may believe travellers, it becomes warmer there than it is in the same latitude on the sea side. Their testimony is strengthened by the vegetables and animals which subsist and multiply there naturally, and do not on our sea coast. Thus Catalpas grow spontaneously on the Missisipi, as far as the latitude of 37°. and reeds as far as 38°. Perroquets [parakeets] even winter on the Sioto, in the 39th degree of latitude. In the summer of 1779, when the thermometer was at 90°. at Monticello, and 96 at Williamsburgh, it was 110°. at Kaskaskia. Perhaps the mountain, which overhangs this village on the North side, may, by its reflexion, have contributed somewhat to produce this heat. The difference of temperature of the air at the sea coast, or on Chesapeak bay, and at the Alleghaney, has not been ascertained; but cotemporary observations, made at Williamsburgh, or in its neighbourhood, and at Monticello, which is on the most eastern ridge of mountains, called the South West, where they are intersected by the Rivanna, have furnished a ratio by which that difference may in some degree be conjectured. These observations make the difference between Williamsburgh and the nearest mountains, at the position before mentioned, to be on an average 6⅛ degrees of Farenheit's thermometer. Some allowance however is to be made for the difference of latitude between these two places, the latter being 38°.8.′17″. which is 52.′22″. North of the former."[15]

In this chapter of *Notes,* Jefferson also included a brief comment on the toleration of humans for extremes of temperature. In an unpublished annotation to his own copy, intended for some future edition, Jefferson added a long explanation of human physiology, the lungs, the performance of the body at high altitude, the newly discovered "vital air" (oxygen), and the best materials for clothing to keep the body warm.

For a complete assessment of climate, more kinds of instruments were needed than thermometers. Soon after acquiring his first thermometer, Jefferson managed to obtain a barometer. In *Notes* he stated that the annual range of barometric pressures did not exceed two inches of mercury and that the pressure at Monticello and Williamsburg differed by 0.784 inches owing to the elevation of Monticello (five hundred feet above sea level). Perhaps the most difficult meteorological feature to measure was humidity because of the unreliability of available hygrometers. This was a disappointment to Jefferson and many others, in no small part because of Buffon's criticism that the climate of the Americas was uniformly unhealthy, being cool and damp.

When Jefferson went to Paris, he quickly came to the conclusion that the French capital was itself a very humid place. Ironically, on that basis he developed a poor opinion of the whole continent. He wrote to Benjamin Vaughan in London (itself a distinctly cloudy place) that "From what I have seen of the climate of Europe, and what I have been able to learn of it from others, it seems to me that it's middle parts are covered by an almost perpetual bank of clouds, extending Northwardly beyond Copenhagen, but not so far as Stockholm, Eastwardly to Switzerland, Southwardly to a little beyond Lyons, and Westwardly to the Western coast of Ireland, and perhaps it is the same which is always hovering over the banks of Newfoundland. Should further enquiry confirm the fact it will be a curious question to examine why the middle parts of Europe should be subject to this general cloudiness while it's Northern and Southern parts enjoy a clear sky, as we do also in America?"[16]

Vaughan replied: "The fact you mention about the clouds in a certain part of western Europe, in a considerable degree corresponds with what I should have supposed. I have turned over some hundred volumes of accounts, antient and modern, and have fully satisfied myself of the *inland and eastern* climates in middle latitudes in the great continents in the Northern hemisphere agreeing with each other, and differing from Western climates in the same situation."[17]

Despite having lived in Paris, and having traveled to the Mediterranean coast of Italy, where "earth and water concur to offer what each has most precious," Jefferson never overlooked an opportunity to express a disdain for things European.[18] Two decades after completing *Notes,* Jefferson warmed to the theme of American superiority of climate in a letter to the Comte de Volney, the French geographer who, as Jefferson knew full well, had lived on both continents. "In no case, perhaps, does habit attach our choice or judgment more than in climate . . . The changes between heat and cold in America, are greater and more frequent . . . than in Europe . . . Habit, however, prevents these from affecting us more that the small changes in Europe affect the European. But he is greatly affected by ours . . . Our sky is always clear; that of Europe always cloudy . . . though we have double the rain, it falls in half the time. Taking all these together, I prefer the climate of the United States . . . I think it a more cheerful one."[19]

Like preceding reports on Virginia by Thomas Harriot and William Byrd, *Notes* was intended in no small part as a promotional piece. As a practical farmer as well as a geographer, and writing for potential immigrants as well as European scholars, Jefferson noted the way in which differences in climate affected the growing conditions for crops. As it turned out, nowhere in America could a finer climate be found than in Virginia. "A more satisfactory estimate of our climate to some, may perhaps be formed, by noting the plants which grow here, subject however to be killed by our severest colds. These are the fig, pomegranate, artichoke, and European walnut. In mild winters, lettuce and endive require no shelter; but generally they need a slight covering. I do not know that the want of long moss, reed, myrtle, swamp laurel, holly and cypress, in the upper country, proceeds from a greater degree of cold, nor that they were ever killed with any degree of cold in the lower country. The aloe lived in Williamsburgh in the open air through the severe winter of 1779, 1780."

One of the most interesting of Jefferson's conclusions about climate, based on both direct measurements and anecdotal information, was that the climate in Virginia, and possibly the rest of America, had become warmer in the previous half-century. "A change in our climate . . . is taking place very sensibly. Both heats and colds are become much more moderate within the memory even of the middle-aged. Snows are less frequent and less deep. They do not often lie, below the mountains, more than one, two, or three days, and very rarely a week. They are remembered to have been formerly frequent, deep, and of long continuance. The elderly inform me the earth used to be covered with snow about three months in every year. The rivers, which then seldom failed to freeze over in the course of the winter, scarcely ever do so now. This change has produced an unfortunate fluctuation between heat and cold, in the spring of the year, which is very fatal to fruits. From the year 1741 to 1769, an interval of twenty-eight years, there was no instance of fruit killed by the frost in the neighbourhood of Monticello. An intense cold, produced by constant snows, kept the buds locked up till the sun could obtain, in the spring of the year, so fixed an ascendency as to dissolve those snows, and protect the buds, during their developement, from every danger of returning cold. The accumulated snows of the winter remaining to be dissolved all together in the spring, produced those overflowings of our rivers, so frequent then, and so rare now."

Jefferson later admitted that the data were never available to make an authoritative statement on the subject of climate change. In 1809 he wrote, "The change which has taken place in our climate, is one of those facts which all men of years are sensible of, and yet none can prove by regular evidence." The year 1772 "is still marked in conversation by the designation of 'the year of the deep snow.' But I know of no regular diaries of the weather very far back."[20] Seven years later he questioned, rather sadly, "May we not hope that the methods invented in later times . . . will at length ascertain this curious fact in physical history."[21] The language he used was, however, always in favor of climate change.

Those who thought the climate had changed mostly agreed upon the possible cause of the putative warming: deforestation of the land.[22] As Benjamin Franklin wrote (tentatively) to Ezra Stiles at Yale College in 1763, "I doubt with you, that Observations have not been made with sufficient accuracy, to ascertain the Truth of the common Opinion, that the Winters in America are grown milder; and yet I cannot but think that in time they may be so. Snow lying on the Earth must contribute to cool and keep cold the Wind blowing over it. When a country is clear'd of Woods, the Sun acts more strongly on the Face of the Earth. It warms the Earth more before Snows fall, and small Snows may often be soon melted by that Warmth. It melts great Snows sooner than could be melted if they were shaded by the trees. And when the snows are gone, the Air moving over the Earth is not much chilled; &c. But whether enough of the Country is yet cleared to produce any sensible Effect, may yet be a question: and I think it would require a regular and steady course of Observations on a Number of Winters in the different parts of the Country . . . to obtain full Satisfaction on the Point."[23]

Others had stated the case more directly. Hugh Williamson, writing in the *Transactions* of the American Philosophical Society, observed in 1770 that "it is not to be dissembled that their winters in Italy were extremely cold about seventeen hundred years ago. Virgil has carefully described the manner in which the cattle are to be sheltered in the winter, lest they should be destroyed by the frost and snow; he also speaks of wine being frozen in the casks."[24] Williamson concluded, however, that it was not the case that as "clearing the country will mitigate the cold of our winters, it will also increase the heat of our summers."

In *Notes,* Jefferson commented on the damage done by intermittent freezing and thawing in early spring. Williamson had discussed that problem in some detail. "A considerable change in the temperature of our seasons may doubtless effect a change in the produce of our lands. Temperate seasons must be friendly to meadow and pasturage, provided we continue to get regular supplies of rain; but of this, there is some reason to doubt, unless our mountains, with which this country happily abounds, should befriend us greatly. The decrease of our frosts and snows in winter, must for many years prove injurious to our wheat and winter's grain. The vicissitudes of freezing and thawing have already become so frequent, that it is high time for the farmer to provide some remedy." Williamson forecast, however, that one day the "tobacco Planter [would] migrate towards the Carolinas and Florida . . . the tender Vine, which would now be destroyed by our winter's frost, in a few years shall supply the North-American with every species of wine . . . Every friend to humanity must rejoice more in the pleasing prospect of the advantages we may gain in the point of Health."

Edmund Augustus Holyoke, who approached the subject as a physician and sought connections between climate and disease, concluded that the greater extremes of heat and cold, compared with temperatures in Europe, were due to the larger number of evergreens growing in America. Thomas Wright, writing in 1794, proposed that, since the effect of deforestation was a reduction in rainfall and humidity, cutting down inland forests would provide a useful way of "drying up the marshes of the maritime parts of North America."[25]

One of the most effective advocates of the deforestation theory was Samuel Williams, onetime Hollis Professor of Natural Philosophy at Harvard and later a minister in Burlington, Vermont. In his *Natural and Civil History of Vermont* (1794), Williams stated that a winter covering of snow prevented loss of the earth's inner heat and kept the roots of plants warm. "When the settlers move into a new township . . . cut down the trees, clear up the lands, and sow them with grain . . . the surface of the earth becomes more warm and dry . . . the cold decreases, the earth and air become more warm; and the whole temperature . . . more equal . . . the number and quantity of the snows decrease." This is what Franklin had suggested to Stiles. But Williams still could not document that such changes had occurred. Instead, he used an extrapolation: "If this computation be admitted, the change of temperature in the winter, at Boston, from the year 1630 to 1788, must have been from ten to twelve degrees."[26]

Amelioration of climate through cultivation of the land was a theme that Buffon had dwelled on lovingly in his *Histoire Naturelle;* it provided him with an explanation of why the Americas were so damp and inhospitable: they were underpopulated and uncultivated. The pervasive influence of Buffon's writings is exemplified yet again in a letter to Jefferson from van Hogendorp. Having received a copy of Jefferson's *Notes,* he wrote, "I conceive that on the borders of the Ocean, perhaps recently freed from the overflowing sea, the low lands of North America, on their first detection by European navigators, may have been more favorable to the breeding of Insects and serpents than of Quadrupeds, as all swampy lands are; but why is nature smaller, or rather less great, in the forming of one animal than of another? And now, after the metamorphosis of Your country, which settlements, as they always do, have rendered wholesome to men, and even dangerous to wild and offensive beasts, now, I say, is there any vestige remaining of Nature's unkindness towards man?"[27]

While it was a commonly held in the latter half of the eighteenth century that the North American climate was warmer than it had been, not everyone agreed. Noah Webster, for example, wrote, "It is a popular opinion that the temperature of the winter season, in northern latitudes, has suffered a material change and become warmer in modern . . . times. This opinion has been adopted and maintained by many writers of reputation . . . indeed I know not whether any person, in this age, has ever questioned the fact . . . It appears to me extremely unphilosophical to suppose any considerable change in the annual heat or cold of a particular country."[28]

As the North American climate continued its usual irregular course of warmer and colder years, acceptance of the warming hypothesis by scientists waned. A more authoritative voice against climate change than Webster's was that of Alexander von Humboldt, the German explorer and scientist whose opinions Jefferson held in high esteem. Humboldt asserted in 1849 that the theories that "destruction of many forests on both sides of the Alleghanys, has rendered the climate more equable . . . are now generally discredited."[29]

Jefferson's friend Chastellux seems to have been the first to suggest to Jefferson yet another interesting phenomenon associated with changing land use. He claimed that because of deforestation, the effect of onshore winds in the afternoons—"sea breezes"—had begun to be felt farther and farther inland. Jefferson repeated this claim, stating in *Notes:* "They formerly did not penetrate far above Williamsburg. They are now frequent at Richmond, and every now and then reach the mountains . . . As the lands become cleared, it is probable they will extend still further westward." Chastellux remarked on the need for a balance in land use. He quoted Franklin to the effect that cutting down the forests helped dry up the marshes; moreover, the "salubrity of the air and even the regular order of the season" depended on the "facility of access which is . . . given to the winds as well as their direction." He therefore concluded that the best way to proceed would be to have scattered clearings and to leave some parcels of forest between the different plantations.[30]

This idea started to catch on. Benjamin Henry Latrobe and William Tatham, writing about growth of the "Sand-Hills of Cape Henry in Virginia" in the American Philosophical Society's *Transactions,* projected that changes in the easterly wind would, "no doubt, when the woods shall be cleared away, blow health and coolness over a portion of lower Virginia, which is now considered as extremely unhealthy." Further, "these easterly winds blowing during the driest and hottest season of the year, carry forward the greatest quantity of sand, and have amassed hills, which now extend about a mile from the beach." He concluded that "if the hills advance at an equal ratio for 20 or 30 years more, they will swallow up the whole swamp, and render the coast a desert indeed, for not a blade of grass finds nutriment upon the land."[31]

The whole question of climate change continued to engross Jefferson after he had completed *Notes.* A great deal of his thinking is revealed in a long letter—really a minor scientific dissertation—on the subject that he wrote to the French physicist Jean Baptiste Le Roy in 1786. The letter contains a remarkably wide-ranging discussion on climate, the effects of land use, ocean circulation, and the Gulf Stream, beginning with the subject of the "sea breezes which prevail in the lower parts of Virginia during the summer months" and the fact that they "had made a sensible progress into the interior country." Jefferson gave a detailed explanation of how the onshore sea breeze was driven by differential heating and cooling between the land and the sea (which produces a corresponding offshore breeze in the mornings) and how deforestation might amplify the effect. Jefferson also asked where the progress of our sea breezes would ultimately be stopped. If cultivation of the land continued westwards, would the easterly breezes "ever pass the Mountaineous country which separate the waters of the Ocean from those of the Mississippi?" He concluded that the mountains, which would remain forested, would always remain a barrier to the sea breezes.

Continuing in this speculative spirit, Jefferson concluded his letter to Le Roy with a little piece of science fiction. Having set out what he knew of oceanic winds and current patterns, he wondered what would happen if the Spanish were to succeed in their dream of cutting a canal through the Isthmus of Panama to shorten, and make infinitely safer, the sea route to the Philippines. "Were they to make an opening . . . however small this opening should be in the beginning, the tropical current, entering with all it's force, would soon widen it sufficiently for it's own passage, and thus complete in a short time that work which otherwise will still employ it for ages . . . The gulph of Mexico, now the most dangerous navigation in the world, on account of it's currents and moveable sands, would become stagnant and safe [but] the gulph stream on the coast of the United States would cease, and with that those derangements of course and reckoning which now impede and endanger the intercourse with those states." He predicted that the fogs on the banks of Newfoundland would disappear, but "it might become problematical what effect changes of pasture and temperature would have on the fisheries."[32] The whole fascinating letter is reproduced as the appendix.

Jefferson's dream was to develop a picture of the climate of the entire United States and, from successive years of observation, to see how it might change. From it one might perhaps develop a full theory of "climate" (and a proof of warming due to deforestation). Though wildly ambitious, the goal was not unreasonable. Franklin had already mapped out the Gulf

Stream, and people were beginning to understand its effects on American and European climates. Within the United States, Franklin had also made the useful observation that the nor'easter storms that pummeled the East Coast, moved across the continent from southwest to northeast, in the opposite direction from their winds. Franklin had come to this discovery by comparing accounts of weather on the same day (October 21, 1743), when many were trying to observe an eclipse of the moon from different parts of the country. The storm that obscured the eclipse in Philadelphia, for example, did not arrive in Boston for another hour or so.

What was needed to derive a full account, let alone an explanatory theory, of the climate of North America was, however, far beyond the observational and computing capacity of the time. Oddly enough, Jefferson did not attempt a full analysis of his own nearly fifty years of data collection at Monticello. In 1817 he made a detailed review of the records from 1810 to 1816 and may have been disappointed (he did not say) to see that there was no change in temperatures during that brief interval. His review essay is, however a snapshot of the climate of that period, covering such details as rainfall and snowfall, the range in flowering times of trees, the appearance of the first peas and strawberries, the arrival of migratory birds, shad in the river, and the first fireflies, and the harvesting of wheat, corn, and peaches. It is a perfect summary of Jefferson's scientific interest in the intersections of climate, geography, nature, and agriculture.[33]

Late in life Jefferson lamented that "of all the departments of science no one seems to have been less advanced for the last hundred years than that of meteorology . . . the phenomena of snow, hail, aurora borealis, looming, etc are as yet very imperfectly understood."[34] Shortly before his death, Jefferson was still optimistic and still held to his prescription of what was needed. He wrote that his *Notes* was "perhaps the first attempt, not to form a theory, but to bring together the few facts then known, and suggest them to public attention . . . and after a few years more of observation and collection of facts, they will doubtless furnish a theory of solid foundation. Years are required for this, steady attention to the thermometer, to the plants growing there, the times of their leafing and flowering, its animal inhabitants, beasts, birds, reptiles and insects; its prevalent winds, quantities of rain and snow, temperature of foundations, and other indexes of climate. We want this indeed for all the States, and the work should be repeated once or twice in a century, to show the effect of clearing and culture towards changes of climate."[35]

CHAPTER FIFTEEN

Redeeming the Wilderness

SHORTLY AFTER RETIRING FROM the presidency, Jefferson was the victim of theft. He had long been yearning to retire to his beloved Monticello to continue the remaking of the house and to build a new one at his Poplar Forest plantation in Bedford County. Also beckoning were numerous intellectual projects, such as planning the University of Virginia and digesting the results of the transcontinental Lewis and Clark expedition.

In *Notes on the State of Virginia,* Jefferson had mused on the origins of the American Indian peoples and their diversification into so many tribes, each with its own language. He suggested that there had been immigration both from the west (from Asia) and from the east (from Greenland). He also believed that "a knowledge of their several languages would be the most certain evidence of their derivation which could be produced."[1] Jefferson had been collecting notes and vocabularies for years. Now he finally would have the leisure to pull all his information together for analysis.

When Jefferson's goods were packed up in Washington, DC, the precious vocabularies were placed in a trunk and treated with great care during the boat trip to Richmond. At Richmond, the James River became impassable to large vessels, and all goods had to be transferred to "bateaus" to be poled and rowed up the Rivanna to Charlottesville. Jefferson's tobacco crops annually made the reverse journey to be transported to Europe.

But in this case some dishonest boatmen, seeing that a chest was marked out for special care, assumed that it contained valuables and stole it. When they opened it and found only papers, they threw them into the river in disgust. Fortunately, a number of the vocabularies were rescued,

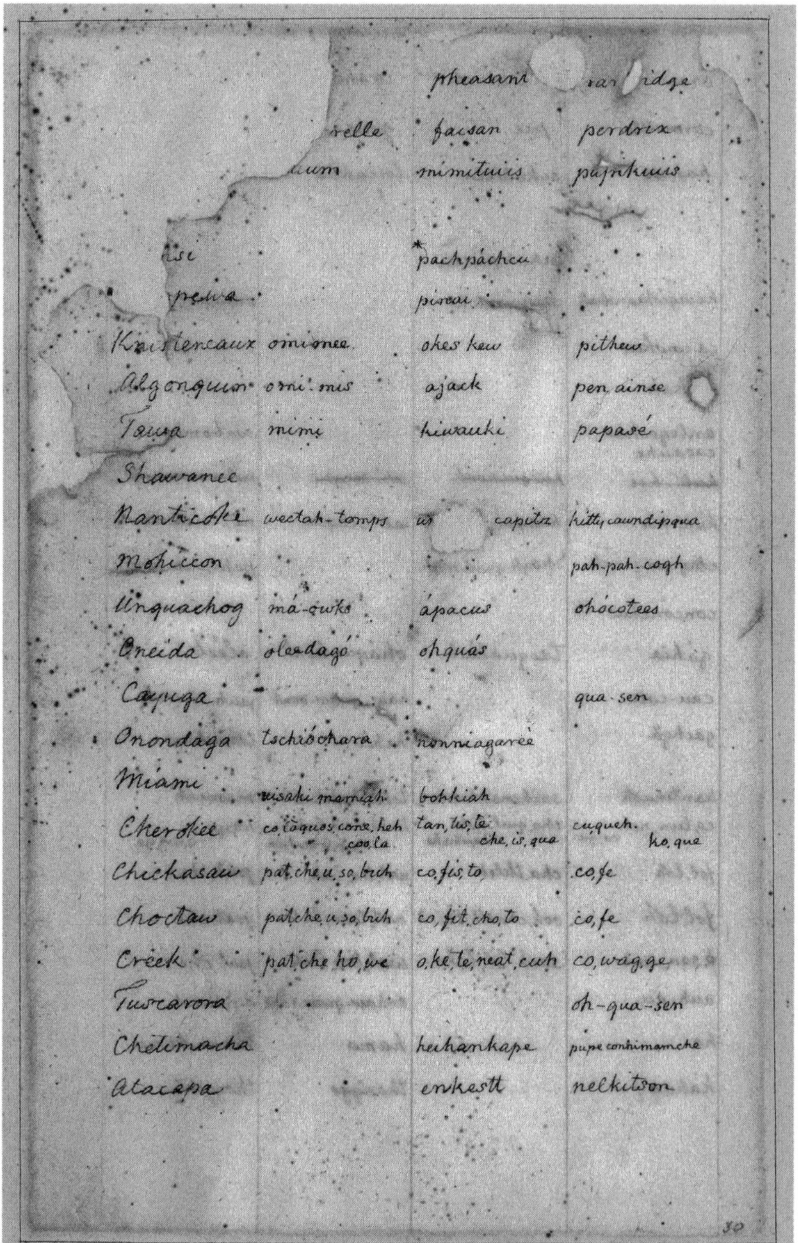

		pheasant	par[illegible]idge
	[illegible]relle	faisan	perdrix
	[illegible]um	mimituiis	pujnkuiis
[illegible]psi		*pachpáchcu	
[illegible]pewa		pireai	
Knistenaux	omimee	okes kew	pithew
Algonquin	omi mis	ajack	pen ainse
Tawa	mimi	kiwauki	papasé
Shawanee			
Nanticoke	weetah-tomps	us[illegible] capitz	kitty cawndipqua
Mohiccon			pah-pah-coqh
Unquachog	má-owks	ápacus	chócotees
Oneida	oleedagó	ohquás	
Cayuga			qua-sen
Onondaga	tschióchara	nonniagarée	
Miami	wisaki mamiah	bohkiah	
Cherokee	co lóquos cone, heh coo la	tan, tis, te che, is, qua	cuqueh. ko, que
Chickasaw	pat che u, so, buh	co, fis, to,	co, fe
Choctaw	patche u, so, buh	co, fit, cho, to	co, fe
Creek	pat che ho, we	o, ke, te, neat, cuh	co, wag, ge
Tuscarora			oh-qua-sen
Chetimacha		heihankape	pupe conhimamche
Atacapa		enkestt	nelkitson

30

This page from one of Jefferson's Indian language vocabularies shows mud stains from having been thrown into a river by thieves. Courtesy of the American Philosophical Society.

and Jefferson gave them to the American Philosophical Society, where they remain still, still showing their mud stains. However, in his heartbreak at the loss, Jefferson gave up his precious Indian language project. In fact, he had become a victim not only of petty larceny but also of geography.

Although the early history of America and Americans was in large part determined by people and events in Europe, it was also defined by the geography of the land itself, with its patchy distribution of ores, its varying soils, its coastal marshes, its rivers—often the best route for travel—and its forests. The eastern part of the country fell into three broad geographical regions: the mountains (the Appalachians, reaching north into the Berkshires and on to the Green and White Mountains of Vermont and New Hampshire), the piedmont ("foot of the mountains") with its low hills and deep soils, and the coastal plain. The eastern mountains are actually a series of folded ridges enclosing deep wide valleys that form a chain of loamy lands—the Great Appalachian Valley—which stretch from Alabama to New York State and then continue up the Hudson and the Connecticut River Valleys. These lands invited early settlers to find access through the mountains from the coastal side.

A line running roughly southwest to northeast, parallel to the mountains, marks the separation between piedmont and coastal plain. This line is not a geographical abstraction like the equator (which, to schoolchildren, famously, is a red line running around the circumference of the earth); rather, it is a geological feature. The "Fall Line" is a low escarpment, in most places measured only in tens of feet in height, between harder (Paleozoic) rocks to the west and the sediments of the coastal plain. Where rivers cross this escarpment they naturally fall through a zone of rapids—hence the name.

As at Richmond, where Jefferson's vocabulary notes were lost, these falls everywhere formed natural barriers to boat traffic. Vessels that came in from the ocean could not pass them; cargoes had to be unloaded, portaged round the falls, and put onto smaller vessels to be taken farther inland. All along the Fall Line, points of portage became settlements, which became centers of commerce and government. The Fall Line is marked now by a string of cities: Tallassee (Alabama); Macon and Columbus (Georgia);

Columbia (South Carolina); Smithfield and Fayetteville (North Carolina); Richmond and Fredericksburg (Virginia); Washington, DC; Baltimore and Conowingo (Maryland); Wilmington (Delaware); Philadelphia (Pennsylvania); Trenton and New Brunswick (New Jersey).

The principal Fall Line runs out into the Atlantic around northern New Jersey, but in New England, cities like Lowell (Massachusetts), Hartford (Connecticut), Albany (New York), Fall River (Massachusetts) and Bangor (Maine) were also located where the navigation of major rivers was interrupted. All these cities were hugely important in the development of the United States. The Fall Line was both a barrier and a source of waterpower; it was stimulus for the creation of centers of population growth, commerce (import-export, transshipping), industry, government and politics, and wealth creation (and with that, poverty). In colonial times, the roads between cities along the Fall Line were bad but well traveled. In modern times, the Fall Line cities have become linked by major highways—U.S. Route 20 (the Upper Boston Post Road) and Interstate 95—that follow the same path.

In addition to the geological Fall Line, a climatological line divided the new country. This was the "sled line." South of Philadelphia, winter snows were ephemeral, neither deep nor dependable enough for the sled to be a useful substitute for the cart in winter. To the north, the sled was essential.

As settlers poured into America, they brought with them European traditions, rebelling against some, retaining many. Governments in the various colonies/states created their laws from classic traditions—for example, by distributing powers among different entities as in England, among crown, church, and people. Laws were mostly based on English common law. As leaders arose, they had their own models—monarchical or republican, agrarian or industrial. Jefferson believed in an ancient Anglo-Saxon style with little government and a solid foundation of yeoman farmers. Architectural styles for buildings large and small were copied and modified from European models. Ships were built according to European styles. Sartorial fashions largely followed those in Europe. Furniture was built from English pattern books until eventually styles diverged as artists and master craftsmen in the New World developed their own styles.

The American wilderness—both the seemingly endless primeval forests and the farmland created once the land had been redeemed by hard work—occupied a unique place in the self-image of the new nation. It came to stand for power, plenitude, and the beauty of the sublime; it nurtured productivity and promised an infinity of natural resources, from iron to gold, from timber to furs. The wilderness was a refuge for enemies (particularly the Indians) and threatening animals. Or it was a reservoir of game for food. Above all, it could be deforested and the wood used for lumber and firewood, or the trees could be burned to fertilize new fields. This wilderness neither looked backward to a European past nor, necessarily, forward to some unknown conditions, and particularly it did not necessarily presage or require a future based on industrialization.

In America it was easy to view nature as the handiwork of a God who had blessed with boundless gifts the new nation. The natural world with which North America had been endowed was in many parts harsh and unforgiving like the God of the Old Testament. It was also bountiful and beautiful; parts of the wilderness were as majestic and sublime as any landscape in Europe. The wilderness inspired awe and reverence, and as Americans pushed west, the number of sublime elements within the wilderness only grew and grew. The vastness of the West was itself metaphorical of the boundless gifts that God had given to America. This view was only strengthened when the colonists discovered that beyond the eastern mountains and forests, there stretched a sea of grass, treeless except for river valleys, all the way to the snow-capped Great Stoney Mountains.[2]

The wildness of America could be seen as a different kind of wilderness; America was a place to be transformed, redeemed, as part of its journey to full nationhood. The Founding Fathers metaphorically led the nation out of a political and social wilderness into a promised land, just as farmers created fertile fields out of the forests. When consulted on the design for a great seal for the United States, Jefferson even suggested the image of the children of Israel being led out of the wilderness (guided by a cloud by day and a pillar of fire at night).[3]

But the American promised land did not flow with milk and honey, free for the taking. It would not even provide the more familiar fruits and vegetables, or forage for cattle, without huge amounts of work. Fields and meadows, orchards and ponds, were needed where at first there were only forests, rivers, and marshes. Roads, which usually followed Indian trails, had to be hacked through the forest to connect farms and settlements and then made safe for travel. Unlike England, America had poisonous snakes and a host of biting insects to be coped with, along with plants like poison ivy, plus bears, wolves, and panthers (and perhaps even the mastodon and the great-claw lurking in some distant lair).

The wilderness in the early days was a barrier to development and a threat. Later it became inspirational instead of foreboding.[4] Even the American Indians, in the later nineteenth century, came to be thought of as romantic innocents. And all of it, land and people, native and immigrant, was now American. In *Letters from an American Farmer,* the superbly named John Hector St. John de Crèvecœur (he began life in France as Michel-Guillaume Saint-Jean de Crèvecœur) described the United States of 1781, rather exaggeratedly, for his European readers: "fair cities, substantial villages, extensive fields, an immense country filled with decent houses, good roads, orchards, meadows, and bridges, where an hundred years ago all was wild, wooded, and uncultivated."[5]

It is hard to conceive how thickly forested eastern North America was when the colonists arrived. As late as 1724, the Reverend Hugh Jones could write of Virginia, "The whole Country is a perfect Forest, except where the Woods are cleared for Plantations, and old Fields, and where have been formerly Indian Towns, and poisoned Fields and Meadows, where the Timber has been burnt down in Fire-Hunting."[6] This forest was more open in the river lowlands, and large areas were cleared land and second-growth forest where the Indians had cultivated and used up the land, then moved on, particularly before contact with Europeans brought disease, decimating their numbers.

One hundred and fifty years after the first European settlers arrived in the New World, when Jefferson sat down to write *Notes on the State of*

Virginia, he could laud the United States as a land of agricultural riches. But for all who had landed in the early days or who later pushed into the western wilds of Massachusetts, New York, Pennsylvania, and Virginia, the constant theme was the contest against the wilderness. The forests had to be cleared, the native wildlife exploited or subdued, the native peoples displaced. European clover and grass species like timothy, rye, and blue grass had to be imported to provide sufficient forage for cattle because the native grasses were unsuitable.[7] New food crops had to be planted. Some, like wheat, peas, and beans, were new to the Americas; some, like corn and squash, were native but new to the immigrants. The wildness of America became a resource for plant hunters looking for new crop plants and for ornamental trees and shrubs to be exported to Europe. Dogwoods, for example, and eventually the majestic California redwoods were America's gift to Europe, and in exchange plantsmen like John Bartram and his son William in Philadelphia introduced a host of Old World plants to the New.[8]

The redeeming of the wilderness, turning it from a hostile environment into an iconic American asset, took a long time. Acre after acre was won by backbreaking work; tree by tree the forest was cut back by hand. The only tools were the ax, the saw, and fire. Like the some of the Indians, a settler would clear the smaller trees by cutting them down and then setting fires that killed the understory of brush. The ashes formed fertilizer. Bigger trees were girdled (a channel was cut deep into the bark, all around the trunk), and the more accessible branches were cut off and piled around the trunk. After a year, when the main tree was dead and dry, the piled-up branches were lit and the tree consumed by the fire.

At first the settlers pushed the Indians back by force or negotiated for the rights to their lands. After the War of Independence, Congress authorized the wholesale development of the lands out of which new "western" states would be created. The money raised from the sale of land went to retire the war debt. As a surveyor and the son of a surveyor, Jefferson knew how important it was for the lands to be mapped properly. "With respect to the sale of our lands, that cannot begin till a considerable portion shall have been surveyed. They cannot begin to survey till the fall of the leaf of this year, nor to sell probably till the ensuing spring. So that it will be yet a twelvemonth before we shall be able to judge of the efficacy of our land office to sink our national debt. It is made a fundamental that the proceeds shall be solely and sacredly applied as a sinking fund to discharge the capital only of the debt."[9]

A committee consisting of Jefferson, Hugh Williamson, David Howell (Rhode Island), Eldridge Gerry (Massachusetts), and Jacob Read (South Carolina) reported to Congress in 1784, recommending that the land be surveyed and divided into a series of new states, each with a gridwork of "sections" of one square mile (640 acres), combined into townships "seven miles square" (later modified to six miles square). The legacy of this committee can be seen in every trip by air across the American Midwest. (Jefferson already had the whole plan in mind before the committee convened, down to names, some improbable, for the new states: "Michigania, Assensipia, Illinoia, Polypotia," and so on.)

The opening up of the West created the potential for a new polarity within the United States: East versus West. As Howell wrote, "The western world opens an amazing prospect as a national fund, in my opinion; it is equal to our debt. As a source of future population and strength, it is a guaranty to our independence. As its inhabitants will be mostly cultivators of the soil, republicanism looks to them as its guardians. When the states on the eastern shores, or Atlantic, shall have become populous, rich and luxurious, and ready to yield their liberties into the hands of a tyrant, the gods of the mountains will save us, for they will be stronger than the gods of the valleys."[10]

Washington, Franklin, and Madison speculated in land beyond the Appalachians. Jefferson did not; he already had all the land he could handle and believed the West should be the province of small farmers. Typically, however, a land speculator with capital available, or using a loan, would purchase forty thousand to fifty thousand acres to subdivide for a profit by selling plots to aspiring farmers. One speculator, Judge Cooper, who opened up lands in western New York State, wrote: "In May 1786 I opened the sales of 40,000 acres, which, in sixteen days, were all taken up by the poorest order of men. I soon established a store, and went to live among them." (As was the case during the California gold rush in the next century, a better living could be made as a store keeper than by manual labor.)

Under the best of conditions (which rarely prevailed), "the poor man, and his class is most numerous, will generally undertake about one hundred acres. The best mode . . . is to grant him the fee-simple by a mortgage on the land . . . He then feels himself, if I may use the phrase, as a man upon record . . . His spirit is enlivened; his industry is quickened . . . he builds himself a barn and a better habitation; plants his fruit-trees, and lays out his garden: he clears away the trees, until they, which were the first obstacles to his improvement, becoming scarcer, become more valuable, and he is at length as anxious to preserve, as he was at first to destroy them . . . He can then can sell at a value resulting from his improvements, and, if so moved, move on and do it again."[11]

Conquering the wilderness was not simply a matter of individual initiative and isolated self-sufficient farms, however. Development of communities, large and small, further and further removed from the big cities, and communication among them all, required roads, and if roads, then laws, and taxes. And therefore government.

Thomas Jefferson and John Adams had strongly differing political views, reflecting not only their education but also geography and the circumstances in which they grew up and, indeed, lived as adults. The philosophies of Jefferson and Adams were a product in part of their attitudes toward the practical and theoretical values of science (natural history and natural philosophy) and in part of their attitudes toward the wilderness, its subjugation, and the exploitation of the land. Jefferson, having grown up in the Virginia piedmont, found himself naturally allied with Philadelphia and the rich farmlands of the southeast. Adams, a product of the hard-scrabble farms and small merchants of Massachusetts found his influences and allegiances in New England.

Different styles of settlement and different patterns of agriculture emerged up and down the new country, and very different economies grew up in the northern and southern states. Virginia and Maryland were hot and suitable for growing tobacco, a commodity of which Europeans could hardly obtain enough. But a climate that was good for tobacco was exhausting for laborers. Rice was found to grow well in the coastal Carolina wetlands, but the heat and the mosquitoes made it lethal. In New England, except in the river valleys, the soils were thin and acid and, once the trees had been cleared and sold or burned for fertilizer, could support only poor crops of grain. New England was far better suited for mixed farming (corn, wheat, hogs, and grazing for cattle) than were Virginia and places farther south. On the northern coasts fisheries and a whaling industry grew up. Pennsylvania remained somewhat in the middle, blessed by good soils and a moderate climate, not as hot in the summer as farther south nor as frigid in the winter as New England.

Slavery was practiced everywhere but was particularly the curse of the more southern states, where whites could not, or would not, work hard continuously in the fields in the summer heat and humidity. At least, free whites would not. Before the number of slaves built up in the early eighteenth century, white indentured servants worked in the fields. As their indentures expired, new labor had to be found. Imported Africans could work in the heat—however unwillingly. Moreover, they were cheap, and there was an economic reason for keeping them so. As Lord Culpepper had written, back in 1683, "In regard of the infinite profit that comes to the king by every black . . . at least six pounds per head per ann[um]. And that the low price of tobacco requires it should be made as cheap as possible, and that blacks can make it cheaper than whites, I conceive it is for his Majesty's interest full much as the country's or rather much more, to have blacks as cheap as possible in Virginia."[12]

Southern plantation culture—where a few lived handsomely on the work of many—differed enormously from the small farm/small business culture of the North, although there were many small farms in Maryland, Virginia, and the Carolinas and some wealthy holders of large estates in the North. Tobacco was farmed as far north as Delaware and Connecticut. A major economic difference between North and South was that planters like Jefferson had to ship their tobacco to England for sale by middlemen and were dependent on a credit system and lived in constant indebtedness. In the North, transactions were more local; they operated on a pay-as-you-go basis and generated cash. Southern plantation owners tended to put their earnings into the acquisition of more slaves and land. In the North, earnings went into investments and the accumulation of monetary wealth rather than real estate. Yankee ingenuity, bred from a shortage of raw materials but an abundance of energy from waterpower, drove manufacturing in the North.

Jefferson, who was acutely aware of the diversity of the American populace in terms of religion, climate, custom, and economy, summed up the differences with a few typically penetrating phrases. In writing to Chastellux, he noted that "my countrymen" are "aristocratical, pompous, clannish, indolent, hospitable . . . I have always thought them so careless of their interests, so thoughtless in their expenses and in all their transactions of business that I had placed it among the vices of their character." By "countrymen" he meant his fellow Virginians, who suffered a "warmth of their climate which unnerves and unmans both body and mind." To this he added a remarkable table:

In the North they are	In the South they are
Cool	Fiery
Sober	Voluptuary
Laborious	Indolent
Persevering	Unsteady
Independent	Independent
jealous of their own liberties, and just to those of others	zealous for their own liberties, but trampling on those of others
Interested	Generous
Chicaning	Candid
superstitious and hypocritical in their religion	without attachment or pretentions to any religion but that of the heart.

Jefferson added, "These characteristics grow weaker and weaker by gradation from North to South and South to North, insomuch that an observing traveller, without the aid of the quadrant may always know his latitude by the character of the people among whom he finds himself. It is in Pennsylvania that the two characters seem to meet and blend and to form a people free from the extremes both of vice and virtue. Peculiar circumstances have given to New York the character which climate would have given had she been placed on the South instead of the North side of Pennsylvania."[13] Jefferson might not have found the differences between regions so easy to caricature if he had compared northerners and southerners holding particular occupations—millers, shopkeepers, ferrymen, or blacksmiths, for example.

Jefferson remained, all his life, suspicious of industrialization and the growth of cities. "You ask what I think on the expediency of encouraging our states to be commercial? Were I to indulge my own theory, I should wish them to practice neither commerce nor navigation, but to stand with respect to Europe precisely on the footing of China. We should thus avoid wars, and all our citizens would be husbandmen. Whenever indeed our numbers should so increase as that our produce would overstock the markets of those nations who should come to seek it, the farmers must either employ the surplus of their time in manufactures, or the surplus of our hands must be employed in manufactures, or in navigation. But that day would, I think be distant, and we should long keep our workmen in Europe, while Europe should be drawing rough materials and even subsistence from America. But this is theory only, and a theory which the servants of America are not at liberty to follow. Our people have a decided taste for navigation and commerce. They take this from their mother country: and their servants are in duty bound to calculate all their measures on this datum: we wish to do it by throwing open all the doors of commerce and knocking off it's shackles. But as this cannot be done for others, unless they will do it to us, and there is no great probability that Europe will do this, I suppose we shall be obliged to adopt a system which may shackle them in our ports as they do us in theirs."[14]

In 1792, Benjamin Smith Barton named a new genus and species of woodland plant for Jefferson, calling it *Jeffersonia diphylla.* "I take the liberty," said Barton, "of making it known to the botanists by the name of Jeffersonia in honour of Thomas Jefferson, Esq. Secretary of State to the United-States . . . In the various departments of this science, but especially in botany and in zoology, the information of this gentleman is equalled by that of few persons in the United-States." Botany was one of Jefferson's strongest and oldest scientific interests—not the simple collecting and naming of plants, or establishing a garden—all of which he did extremely well—but experimentally.

At first, Jefferson did not have an elaborate flower garden at Monticello. All his efforts went into a vast vegetable garden. Jefferson's "final version" of Monticello included a pleasure garden on the south side of the house in which he adopted curving lines and shapes, abandoning traditional classical straight lines and right angles. These decorative elements contrasted markedly with the rectangular precision that he maintained when laying out the more functional vegetable gardens and orchards that echoed the ordered, numbered, vegetable plots of his father and mother's garden at Shadwell.

Even before any foundations had been started for the great house at Monticello, Jefferson marked out and had the slaves dig an enormous oblong vegetable garden between Mulberry Row (the long lane of mulberry trees with its quarters for the house slaves and workers) to the north and the orchard below. It formed one huge "hanging garden" or terrace, supported on the downhill side by a stone wall against which he grew his favorite "Marseilles" figs. Halfway along the wall he built a raised pavilion where he could sit and read, and survey his lands all the way over to another farm, Tufton, that he had inherited from his father. The garden started out as one 668 feet long and 80 feet wide; Jefferson later had it extended to a rectangle 1,000 feet long, encompassing two acres, intending to subdivide it into twenty-four neat rectangular beds.

Plantings in the twenty-four square plots were to be arranged in three groups. At the western end, were the "fruits" including beds of asparagus, peas (Leadman's dwarf), snap beans, haricot beans, cucumbers, melons, peppers, "tomatas," okra, artichokes, and squashes. Root vegetables like carrots, salsify, beets, garlic, leeks, and onion were in the middle beds. At the eastern end were the leaf vegetables: scallions, nasturtiums, lettuce, endive, "terragon," celery, spinach, sorrel, mustard, sea kale, cauliflower, broccoli, and cabbages. Beans and black-eyed peas were also a field crop in the farm, as were several different kinds of potatoes (long, round, Indian, Irish).[15]

Beyond the vegetable garden were Jefferson's plantings of grapes and soft fruits—strawberries, raspberries, and gooseberries—again in neat rows. Farther down the sunny slope were the orchards of fruit trees in strictly regimented rows of cherries, peaches, apples, persimmons, and pecans. (Jefferson constantly promoted the virtues of the pecan tree and was one of the first to recognize it as a distinct species.) Jefferson did at least some of the grafting of fruit tree stock himself. He was much intrigued by his peach trees and produced one of his famous sums concerning the value of their annual prunings (peaches bear only on new wood). He asked his son-in-law Thomas Mann Randolph to make measurements on the use of the peach trees as a source of firewood. "Your experiment . . . proves my speculation practicable, as it shews that 5. acres of peach trees at 21. feet apart will furnish dead wood enough to supply a fireplace through the winter, & may be kept up at the trouble of only planting abut 70. peach stones a year. Suppose this extended to 10. fireplaces, it comes to 50. acres of ground, 5000 trees, and the replacing of about 700 of them annual."[16]

In this vegetable garden Jefferson experimented with every variety of seed he could get his hands on, acquiring stock by purchase, gift, or exchange with other gardeners around the world. In Paris he had developed a close association with Madame de Tessé, herself a keen gardener, who encouraged him to plant flowers around the Hôtel de Langeac. He arranged for packages of American seeds to be sent to her, and they continued to exchange plants and seeds long after Jefferson had returned to America. When he was president, he wrote to her that "I rarely ever planted a flower in my life . . . I believe I shall 'become a florist.'"[17] André Thouin, superintendent of the royal Jardin des Plantes became a friend with whom he exchanged materials for nearly twenty-five years. At home, he developed close relationships with the prominent commercial plantsmen in Philadelphia—men like John and William Bartram (really the founders of the whole import-export enterprise with Europe) and Bernard McMahon—as well as fellow landed "gentry," such as William Hamilton, with his marvelous gardens at the Woodlands.

Jefferson's order to William Prince at Flushing, Long Island, in 1791, is an example of his purchases: "Sugar maples. All you have. bush cranberries. All you have. 3. balsam poplars. 6. Venetian sumachs. 12. Burée pears. 6. Brignola plumbs. 12. apricots. I leave to you to fix on three or four of the best kinds, making in the whole 12 trees. 6. red Roman nectarines. 6. yellow Roman nectarines. 6. green nutmeg peaches. 6. large yellow clingstone peaches ripening Oct. 15. 12. Spitzenberg apples. I leave to you to decide on the best kind, as I would chuse to have only one kind. 6. of the very earliest apples you have. Roses. Moss Provence. Yellow. Rosa mundi. Large Provence. The monthly. The white damask. The primrose. Musk rose. Cinnamon rose. Thornless rose. 3 of each, making in all 30. 3. Hemlock spruce firs. 3. large silver firs. 3. balm of Gilead firs. 6 monthly honey suckles. 3 Carolina kidney bean trees with purple flowers. 3. balsam of Peru. 6. yellow willows. 6. Rhododendrons. 12. Madeira nuts."[18]

He field-tested seeds brought back from the West by Lewis and Clark as well as seeds from all over Europe. Detailed entries in his Garden and Farm Books show that Jefferson experimented over the years with at least 330 different varieties of food plants in his huge garden, always seeking to improve taste, hardiness, and productivity. He tried forty kinds of beans, fifteen of corn, seventeen of lettuce, and an amazing fifty-four varieties of peas. In his orchards he field-tested soft fruits like raspberries and fruit trees. He tried out Italian rice in his fields and trees like olives and sugar maples on the hillsides. "I am curious to select only one or two of the best species or variety of every garden vegetable."[19]

In experimenting, Jefferson was acting as a scientist, a farmer, and a patriot. With the Embargo Act of 1807 and increasing tension between the United States and Europe, it was important to be independent from Europe. He wanted "to make every thing we want within ourselves and have as little intercourse as possible with Europe."[20]

Perhaps ironically, having drastically modified his mountaintop and created a formal vegetable garden, Jefferson attempted to create on the estate one area that was to be a romantic re-creation of nature—a "Grove" —by cutting out "superabundant plants" and opening up vistas. Here, he noted to himself, he would "thin the trees, cut out stumps . . . cover the whole with grass. Intersperse Jessamine, honeysuckle, sweetbriar, and even hardy flowers which may not require attention. Keep it in deer, rabbits,

Peacocks, Guinea poultry, pigeons etc. let it be an asylum for hares, squirrels, pheasants, partridges, and every other wild animal (except those of prey) . . . procure a buck-elk, to be, as it were, monarch of the wood; but keep him shy . . . a buffalo might be confined also . . . [and arrange] benches or seats of rock or turf."[21]

The Grove proved impractical, but Jefferson did establish a deer park. His slave Isaac Jefferson noted that it was "two or three miles round and fenced in with a high fence, twelve rails double-straked and ridered." The deer were fed at sunup and sundown: he "called me up and fed 'em wid corn . . . gave 'em salt, got right gentle; come up and eat out of your hand." If Jefferson heard hunters near the park, he "used to go down thar wid his gun and order 'em out."[22]

Elk and wolves had already become extinct in Virginia in Jefferson's time, and deer and coyotes were now a pest, which tells us something about the consequences of "redeeming" the wilderness, of making it suitable for human habitation and using it for farming. The new agricultural regimes, with their imported plant species and intense cropping, turned out to be unstable.

As a southern farmer, Jefferson was hit hard by the War of Independence, which closed English markets to his principal crops. Even before that, a more fundamental problem had arisen: yields were dropping because growing tobacco seriously depleted the soil of nutriments. The redemption of the wilderness—its being deforested and made into fertile fields—had started to fail as a path to fruitful cultivation. In 1759 the Virginia Assembly had to set up a committee to investigate ways of diversifying agriculture beyond the cultivation of tobacco.

Washington tried spreading mud from the bottom of the "Potomak" onto his fields. Elsewhere, marl, ashes, lime, and seaweed (especially in New England) were used to put nutrients back into the soil. In the early days, the solution to depletion of the soil was simple—one did what the agriculturalists in an apparently limitless wilderness had done from time immemorial: move on. As Jefferson wrote to George Washington in 1793, "We can buy an acre of new land cheaper than we can manure an old one."

On tobacco plantations, there was, however, less animal manure to spread on the fields than there was in a mixed-use northern farm. Now, with larger, more permanent settlements, fertilizers were sought, or new crops were.

Jefferson noted that "[in] the highland where I live . . . the culture was tobacco and Indian corn as long as they would bring enough to pay labor . . . after four or five years rest they would bring good corn again, and in double that time perhaps good tobacco. Then they would be exhausted by a second series of tobacco and corn."[23] Eventually the smaller plantation owners, like Jefferson and Washington, and all the great plantation owners, like the Carters on the Northern Neck of Virginia, realized that the monoculture of tobacco had to be replaced by some equally valuable but less damaging crop or crops, and the soil had to be conserved.

This was an ideal problem for Jefferson, no longer occupied with affairs of state, to turn his mind to. The obvious crop to replace at least some of the tobacco acreage was wheat, for which Europe had plenty of need, but wheat was susceptible to rusts (viruses) and insect pests. Indian corn (maize) was also valuable, but it tended to deplete the soil almost as fast as tobacco did. Many farmers alternated wheat and corn in their fields, but that practice also soon exhausted the upland piedmont soils. Jefferson kept up with tobacco as his main crop until 1793 and then experimented with wheat. But a more scientific solution to the problem was needed.

In 1786, George Washington began a correspondence with the English political and agricultural economist Arthur Young, hoping to solve some of his problems, which he expressed forthrightly. "The system of Agriculture (if the epithet of system can be applied to it) which is in use in this part of the United States, is as unproductive to the practitioners as it is ruinous to the landholders. Yet it is pertinaciously adhered to. To forsake it; to pursue a course of husbandry which is altogether different & new to the gazing multitude, ever averse to novelty in matters of this sort, & much attached to their old customs, requires resolution; and without a good practical guide, may be dangerous, because of the many volumes which have been written on this subject, few of them are founded on experimental knowledge—are verbose, contradictory, & bewildering."[24]

Jefferson followed developments in European farming himself, and like Washington, he began to adopt the system of crop rotation that was being promoted by Young.[25] The principle lay in planting crops that increase soil fertility during the fallow period of the planting cycle. The plants could be plowed in to decompose and enrich the soil. As we now know, bacteria associated with the roots of legume crops, such as clover and peas, take nitrogen from the atmosphere and release it into the soil, converted into a form that plants can use, making them a particularly beneficial crop to plant. Noah Webster, who had also become an advocate of "green manure," wrote Washington a long nagging letter about it in 1790: "I believe, sir, fertility may be restored to exhausted fields . . . by seeding them with some succulent vegetables, as oats, peas, beans, buckwheat, & turning in the full growth or suffering the whole to rot upon the land."[26]

Jefferson launched into crop rotation schemes, devising plans for dividing his farms into manageable units and alternating crops. His most elaborate scheme was an eight-year cycle for eight fields rotating in turn: "wheat and [fall] fallow; peas and corn thinly interplanted; wheat and [fall] fallow; potatoes and corn interplanted; rye and [fall] fallow; clover; clover; clover."[27] As with so many of Jefferson's paper calculations, this one was impractical. He also dreamed up a six-year plan, but by 1798 he had put in practice a simple three-year rotation in the European style: "one year of wheat and two of clover in the stronger fields, or two of peas in the weaker, with a crop of Indian corn and potatoes between every other rotation . . . Under this easy course of culture, aided with some manure, I hope my fields will recover their pristine fertility."[28]

In the hilly country of northern Virginia, another problem was soil conservation. A British visitor to Monticello in 1807 observed, "The President was considered a very bad farmer. He had some excellent red land . . . but his estate suffered prodigiously from the tendency of the soil to gully, which cannot well be prevented on hilly ground, and is very much promoted by the culture of Indian corn."[29] Jefferson adopted the practice of contour plowing, which his son-in-law Thomas Mann Randolph also championed. The result was improved soil conservation, which was particularly important on the steep slopes around Monticello. And, of course, Jefferson invented an improved design for a plow that turned the soil over more deeply and brought more nutrients to the surface than the ordinary plow did.

It did not take long for introduced crop plants like wheat to fall to pests. The possible strategies to deal with insect infestations were limited. Landon Carter, at his plantation in Sabine Hall, Virginia, had some success against "fly-weevil" by carefully observing the cycle of infestation and working out how to break it. In a classic of early American experimental science, he increased the heat of the grain in storage and reduced the population of overwintering insects.[30] Then a new problem appeared in the form of the Hessian fly, so named because it was supposed to have been introduced by Hessian mercenaries fighting for the British. (Later studies suggested that it in fact had come from Asia and had been seen in America as early as 1768.)

In 1791, when the British refused to accept imports of "contaminated" American wheat, Jefferson brought together a committee of the American Philosophical Society to investigate the situation. The members included Benjamin Rush, Caspar Wistar, Charles Thomson, and Benjamin Barton Smith. This was the sort of challenge to appeal to Jefferson. The solution seemed to lie first in gathering as much information as possible about the natural history of the fly, its life cycle, and the pattern of its infestation of wheat. The committee made up a questionnaire for gathering information. When did the eruptions of insects occur? What part of the plant did they attack? Were other crops affected? Did manuring help? Were some varieties of wheat less susceptible than others? What effect did cold have? Did other animals eat the insect? What farming practices affected the insect?[31]

The same year, Jefferson and his friend James Madison made a "northern journey" through New York and New England, one purpose of which was to see the problem in person. (They were also trying to cement old political alliances and make new ones with a view to the next presidential election.) Starting in New York City, they traveled to Lake George and Lake Champlain, where they were impressed by the dramatic landscape, the abundant fishes, and the "musketoes & gnats, & 2 kinds of biting flies." They headed back through Vermont, down to Connecticut, and across to Long Island, covering some nine hundred miles in their month of travel.

Jefferson, as on his continental tours, kept a lookout for interesting and useful plants—the sugar maple, for example, which he realized could make America less dependent on imported sugar. The two travelers probed into different farming practices and inspected new inventions, viewing factories for processing barrels of fish, machines for drawing water from wells, and a nailery, where Jefferson took notes for his own nail workshop. On Long Island, Jefferson unexpectedly managed to add to his data on American Indian languages. He jotted down on the back of an envelope a vocabulary of some two hundred words from some women of the Algonkian Unquachog people.

The outbreak of the Hessian fly had been particularly severe on Long Island. From farmers there, Jefferson learned the outlines of its life cycle. It lays its eggs between the leaves and wheat stem of young plants. When the maggots hatch, they suck the juices of the stem, stunting and weakening the plant. Even if the plant does not fall over, seed production is reduced. In their next stage of life, the maggots pupate, fall to the ground, and hatch out. The mature flies live in the wheat stubble until the new crop of wheat germinates. They lay eggs, and the cycle starts again. Eventually Jefferson was sent batches of eggs to examine, and he observed the life cycle under his own microscope.

Jefferson petitioned the British to lift the embargo on the grounds that the wheat grains themselves did not contain eggs or maggots, but to no avail. In fact, the whole affair petered out without resolution and the American Philosophical Society Committee, unable to discover any simple way to eradicate the fly, made no formal report. All that was available to the farmer was a recommendation to burn the stubble fields after harvest and sow as late as possible in the year. Various varieties of "bearded wheat" with tough stalks were found to be the most resistant to infection. Jefferson's informants said that manuring the fields also helped by producing a sturdier stalk and a stronger plant.

The flies spread unchecked across the Hudson and Delaware Rivers, circled Chesapeake Bay, and attacked George Washington's wheat fields at Mount Vernon. Monticello had its first infestation around 1811. The insects' numbers flourished erratically: there were good years and bad. Of his crop in 1817, Jefferson wrote, "We . . . must be contented with so much . . . as that miserable insect will leave us. This remnant will scarcely feed us the present year, for such swarms of the Wheat fly were never before seen in this country."[32] The Hessian fly is still a major pest of American wheat. One of the more resistant varieties of modern bearded wheat is named for Jefferson.

Geography really mattered for the young United States. If Jefferson's committee had found a solution to the Hessian fly problem, neither it nor the federal government could have imposed new practices on all the nation's farmers. Meanwhile, failure to control the fly in one area had consequences in other farms, counties, or states. The situation is emblematic of the central, far more consequential problems already facing the new republic. To what extent would the new nation act, and be governed, as a single unit, and where did the independence of each state lie? How strong should the central government be, and how broad its reach? These questions continued to pit the Republican-Democrat Jefferson against his Federalist opponents.

Meanwhile, science and common sense helped Jefferson learn to manage his farmlands better, but they could not keep him solvent. He lived too well, and was too generous—too quick to secure a loan for a friend or to take one out for himself. He started out with a burden of debt from his wife's family, and in the end, the bounty of the American landscape, redeemed but already compromised, was not sufficient to support him. He died very much a Virginian and deeply in debt. The bounty of his intellectual vision, however, lives on, and nowhere more splendidly than in his ambitious vision for the ultimate geographical scope of the United States, expanded into the western wilderness.

CHAPTER SIXTEEN

The Unknown West

WITH THE POSTING TO PARIS, six weeks spent traveling in southern England, the three-month tour of France and Italy, the six-week tour of Holland and the Rhineland, and the monthlong northern tour of the United States, Thomas Jefferson was a well-traveled man, especially for someone who craved (or claimed to crave) the solitude of life at home. His most important voyage of discovery, which began the process of cementing together the modern geographic vision of the country, was conducted vicariously. The journey of the Corps of Discovery, made between 1803 and 1806 under the leadership of Meriwether Lewis and William Clark, was in many ways a reflection of Jefferson himself, the man who commissioned and planned it. The expedition was one of his greatest scientific achievements.

The western wilderness extended, as everyone knew, two thousand miles or more to the Pacific Ocean. Most of it was unknown (except, of course, to its Indian inhabitants). Explorers from the Pacific side had not penetrated far inland before they saw a barrier before them, range after range of the magnificent Great Stoney Mountains, but some also told of a mighty river in the north (the Columbia) that afforded access to the hinterland. In 1800, French and Spanish traders, trappers, and missionaries knew a great deal more about these western lands than did the English speakers who ruled from the Appalachians eastward. But which group of Europeans would finally discover the secrets of the western wilderness? Indeed, who would colonize, exploit, and own it?

Jefferson the statesman saw the vast spread of land from the west bank of the Mississippi to the Pacific as a political, economic, and scientific opportunity. The French and Indian War had given British America the land reaching to the east bank of the Mississippi, including the present states of Wisconsin, Illinois, Tennessee, and Mississippi. The Louisiana Purchase of 1803 (negotiated during Jefferson's presidency by Robert Livingston and James Monroe for the United States and by none other than Barbé-Marbois for France) had doubled the size of the United States and extended its western reach (to a line eventually to be settled by negotiation with the French and Spanish as the Line of 1819). The new territory included what are now Kansas, eastern Colorado, Nebraska, and most of Wyoming and Montana, but not western Montana, Idaho, Oregon (which remained disputed with Britain and France until 1846), Utah, Nevada, California, Arizona, New Mexico, or Texas (which remained with Spain until 1848). The huge purchase made possible the dream of spanning the continent from Atlantic to Pacific, controlling a sweep of land free of French or Spanish interference.

Like the eastern wilderness, these western lands had to be placed under the control of the United States government. The forests and plains could not be cleared for American farmers, the rivers made safe for travel, and roads thrust to the west until purchases and treaties had been made with the Indians. And before that, an inherent right of dominion of these parts of the American continent had to be established. The French had not sold the title to the western lands. Instead, "the United States bought the Discovery rights to a limited sovereignty over the territory, the right to be the only government the Indian Nations could deal with politically and commercially, and the preemption right, the exclusive option to purchase the real estate whenever the owners, the Indian Nations, chose to sell."[1] Over the next 150 years, three hundred million dollars had to be added to the cost of the Louisiana Purchase to buy Indian lands.

One mission of the Corps of Discovery was therefore to make treaties with Indians along the route up the Missouri and across the mountains to the Pacific. Lewis and Clark were both to recognize the Indians' sovereignty over their own lands and, in a curious piece of diplomatic nicety, insist that the government in Washington, DC, was now the Indians' new "father." The Indian tribes were sovereign nations, but they were to be told that they were also subject to the overarching laws and policies of the

United States. Making treaties with them was essential; the alternative was the use of force. As Jefferson put it, "The Indians can be kept in order only by commerce or war. The former is the cheapest."[2]

The mission of the Lewis and Clark Expedition was complex. Above all, they were to plant the American flag at the mouth of the Columbia River and to claim its whole watershed for the United States. In the process, they were to find a navigable route from the Missouri to the Pacific. They would explore, map, collect new animals and plants, find useful minerals and ores, and dig out fossils. They would establish trading relations with the Indian tribes and begin the process of making treaties for their land.

Lewis and Clark were sent to Philadelphia for training by the country's best experts. Benjamin Smith Barton at the University of Pennsylvania, America's first professor of botany, taught Lewis how to collect and preserve live plants, pressed specimens, and seeds. Dr. Benjamin Rush gave them both instructions in field medicine. Dr. Caspar Wistar instructed Lewis on fossils and geology and showed him how to put up skins and skeletons. Jefferson added his own instructions for preserving bird skins.[3] Lewis went to David Rittenhouse in Philadelphia and the mathematician Andrew Ellicott in Lancaster, Pennsylvania, for lessons in making astronomical observations and surveying. Robert Patterson at the University of Pennsylvania (Ellicott's teacher) also helped with surveying methods although Lewis may not have had the full necessary set of skills in mathematics.

Jefferson wrote a long letter to Lewis in June 1803 confirming the details of the mission as authorized by Congress. "The object of your mission is to explore the Missouri river, & such principal stream of it, as, by it's course & communication with the waters of the Pacific Ocean, whether the Columbia, Oregan, Colorado or and other river may offer the most direct & practicable water communication across this continent, for the purposes of commerce."[4]

Finding a navigable passage from the Atlantic to the Pacific (the fabled Northwest Passage) without rounding Cape Horn, had been a dream of American and European powers since at least Elizabethan times. Hopes of a northern route through the Arctic Ocean were constantly dashed, although attempts to find one were still made throughout the nineteenth century. When Roald Amundsen eventually found a way through in 1905, the route was not commercially viable.

Jefferson pinned his hopes once again on travelers' tales. Explorers, traders, and fur trappers pushing across the continent often brought back accounts of a great river emptying into the Pacific in the region of modern Oregon. Traced north and west, the Missouri seemed to continue unabated across the American interior, so it seemed possible that the watersheds of the two might be accessible one to the other, although since the watersheds led in opposite directions, there had to be an elevated "divide" between the two. But that divide might be crossed with portages. Perhaps.

Jefferson reassured Lewis. "Your mission has been communicated to the ministers here from France, Spain & Great Britain, and through them to their governments; & such assurances given them as to its objects, as we trust will satisfy them." This was important because the Lewis and Clark Expedition was not the first attempt to survey the western reaches of the continent, and previous ventures had been met with deep suspicion by the French and Spanish.

The terms of the expedition required detailed attention to accurate mapping and extensive observation of everything from geology and hydrology to natural history. Jefferson's instructions were comprehensive. "Beginning at the mouth of the Missouri, you will take *careful* observations of latitude & longitude, at all remarkeable points on the river, & especially at the mouths of rivers, at rapids, at islands, & other places & objects distinguished by such natural marks & characters of a durable kind, as that they may with certainty be recognised hereafter . . . The variations of the compass too, in different places, should be noticed . . .

"Altho' your route will be along the channel of the Missouri, yet you will endeavor to inform yourself, by enquiry, of the character & extent of the country watered by it's branches, & especially on it's Southern side. The North river or Rio Bravo which runs into the gulph of Mexico, and the North river, or Rio colorado which runs into the gulph of California, are understood to be the principal streams heading opposite to the waters of the Missouri, and running Southwardly. Whether the dividing grounds between the Missouri & them are mountains or flatlands, what are their distance from the Missouri, the character of the intermediate country, & the people inhabiting it, are worthy of particular enquiry. The Northern waters of the Missouri are less to be enquired after, becaue they have been ascertained to a considerable degree, & are still in a course of ascertainment by English traders, and travellers. But if you can learn any thing certain of the most Northern source of the Missisipi, & of its position relatively to the lake of the woods, it will be interesting to us . . .

"As far up the Missouri as the white settlements extend, an intercourse will probably be found to exist between them & the Spanish post of St. Louis opposite Cahokia, or Ste. Genevieve opposite Kaskaskia. From still further up the river, the traders may furnish a conveyance for letters. Beyond that, you may perhaps be able to engage Indian to bring letters for the government to Cahokia or Kaskaskia, on promising that they shall there receive such special compensation as you shall have stipulated with them. Avail yourself of these means to communicate to us, at seasonable intervals, a copy of your journal, notes & observations, of every kind, putting into cypher whatever might do injury if betrayed.

"Should you reach the Pacific ocean inform yourself of the circumstances which may decide whether the furs of those parts may not be collected as advantageously at the head of the Missouri (convenient as it supposed to the waters of the Colorado & Oregan or Columbia) as at Nootka sound, or any other point of that coast; and that trade be consequently conducted through the Missouri & U.S. more beneficially than by the circumnavigation now practised." Jefferson added: "Some account too of the path of the Canadian traders from the Missisipi, at the mouth of the Ouisconsin to where it strikes the Missouri, & of the soil and rivers in it's course, is desireable."

Also, "on your arrival on that coast endeavor to learn if there be any port within your reach frequented by the sea-vessels of any nation, & to send two of your trusty people back by sea, in such way as shall appear practicable, with a copy of your notes: and should you be of opinion that the return of your party by the way they went will be eminently dangerous, then ship the whole, & return by sea, by way either of cape Horn, or the cape of good Hope, as you shall be able."

All along the route the Corps of Discovery followed in the footsteps of Spanish and French predecessors. The party did not encounter Indians who had never seen a white person before. On the Pacific coast they even met with Indians who had a smattering of English learned from sailors—and an amazing vocabulary of swear words. They sent back maps that changed the public perception of the continent, as well as seeds, Indian artifacts, live and preserved animals, and notebooks full of detailed geological, hydrological, and topographic information. They set in place land agreements and opened the door for the development of many others.

The expedition took along the standard surveying tools: two sextants, a theodolite, a level (artificial horizon), a surveyor's compass, and a chain measuring two poles in length (one-twentieth of a furlong). Their luxury item was a hugely expensive chronometer, made by Thomas Parker in Philadelphia for $250, which was necessary for determination of longitude. Recent studies show that their surveying, mapping, and hydrological work was remarkably accurate.[5] Ellicott, Rittenhouse, and Patterson had trained them well.

In geology they sent back samples of silver ores from Mexico given them by the Osage Indians and a small number of mineral specimens collected along the route.[6] They found fossil woods, shells, and a large bone, none of which survives. They collected a "Jawbone of a fish or some other animal found in a cavern a few miles distant from the Missouri." This specimen, from present-day Harrison County, Iowa, survived the journey home and is now named *Saurocephalus lanciformis.*[7]

To woo the Indians along the route and persuade them of the superiority of their new "white fathers," the expedition took along a huge inventory of trade goods. These included the much-favored blue beads, but they did not take enough.

Jefferson had written: "The commerce which may be carried on with the people inhabiting the line you will pursue, renders a knolege of those people important. You will therefore endeavor to make yourself acquainted, as far as a diligent pursuit of your journey shall admit, with the names of the nations & their numbers; the extent & limits of their possessions; their relations with other tribes of nations; their language, traditions, monuments; their ordinary occupations in agriculture, fishing, hunting, war, arts, & the implements for these; their food, clothing, & domestic accommodations; the diseases prevalent among them, & the remedies they use; moral & physical circumstances which distinguish them from the tribes we know; peculiarities in their laws, customs & dispositions; and articles of commerce they may need or furnish, & to what extent . . .

"In all your intercourse with the natives, treat them in the most friendly & conciliatory manner which their own conduct will admit; allay all jealousies as to the object of your journey, satisfy them of its innocence, make them acquainted with the position, extent, character, peaceable & commercial dispositions of the U.S. of our wish to be neighborly, friendly & useful to them, & of our dispositions to a commercial intercourse with them . . . Carry with you some matter of the kinepox; inform those of them with whom you may be, of it'[s] efficacy as a preservative from the small-pox; & instruct & incourage them in the use of it. This may be especially done wherever you winter."

Jefferson had high hopes that the expedition would be able to gather data for his idea that Indian languages could be analyzed to reveal the relationship of Indian languages to one another and the origins of the Native American peoples. "The late discoveries of Captain Cook, coasting from Kamschatka to California, have proved that, if the two continents of Asia and America be separated at all, it is only by a narrow streight. So that from this side also, inhabitants may have passed into America: and the resemblance between the Indians of America and the Eastern inhabitants of Asia, would induce us to conjecture, that the former are the descendants of the latter, or the latter of the former: excepting indeed the Eskimaux, who, from the same circumstance of resemblance, and from identity of language, must be derived from the Groenlanders, and these probably from some of the northern parts of the old continent. A knowledge of their several languages would be the most certain evidence of their derivation which could be produced."[8]

The expedition took along standardized forms with a list of 315 terms, beginning with "Fire, water, earth, air" and ending with "cowardly, wife, foolish, I, you, he." At least fourteen of the "word lists" were completed. Added to the forty or so vocabularies that Jefferson had collected himself over the years, they formed the priceless collection, part of which was lost at Richmond.

Today, most public attention is paid to the results of the expedition in terms of natural history. As Jefferson wrote, "Other objects worthy of notice will be the soil & face of the country, it's growth & vegetable productions, especially those not of the U.S. the animals of the country generally, & especially those not known in the U.S. the remains or accounts of any which may be deemed rare or extinct; the mineral productions of every kind; but more particularly metals, limestone, pit coal, & saltpetre; salines & mineral waters, noting the temperature of the last, & such circumstances as may indicate their character; volcanic appearances; climate, as characterized by the thermometer, by the proportion of rainy, cloudy, & clear days, by lightening, hail, snow, ice, by the access & recess of frost, by the winds prevailing at different seasons, the dates at which particular plants put forth or lose their flower, or leaf, times of appearance of particular birds, reptiles or insects."

We cannot give Jefferson direct credit for the species of plants and animals that were discovered in the West by the expedition. But it was because of Jefferson's insistence on the scientific goals of the expedition that species new to European and American science were known so soon and that many species of useful and ornamental plants were introduced not only to the eastern United States but also to Europe.[9]

Besides collecting seeds and live plants, Lewis and Clark brought back many specimens dried as herbarium sheets, as Lewis had been shown how to do. Of the dried specimens, one per sheet of paper, more than 200 survive.[10] Of the approximately 178 new kinds of plants collected by Lewis and Clark, some 80 were species new to European science (the rest were subspecies: botanists tend to argue about what a species is). Familiar plants in the list include Osage orange (the wood of which Indians used for bows), purple coneflower, prickly pear, western paper birch, bitterroot, two species of cottonwoods, cut-leaf daisy, Jacob's ladder, red-flowering currant, golden currant, and sage brush, Pawnee and Arikara corn, Arikara bean, Mandan tobacco, gooseberry, and wild salsify. Most of the forest trees were new. Among the ornamental plants were a yellow fritillary, the Oregon grape, and the lovely snowberry, a bush that is covered with small white berries in the fall.

Among the new animals were Pacific sturgeon and steelhead and cutthroat trout. They saw such now-familiar birds as Clark's nutcracker, Lewis's woodpecker, the white-fronted goose, Franklin's grouse, the sage grouse, the northern flicker, the black-billed magpie, the western tanager, and the broad-tailed hummingbird. Of mammals, they made the first known records, in English at least, of the pronghorn, the plains prairie dog, the mule deer, the sea otter, the grizzly bear, the coyote, the gray wolf, the Rocky Mountain bighorn sheep, and the western badger.[11]

On April 7, 1805, Lewis sent down the Missouri from Fort Mandan (North Dakota) a shipment of twenty-six packages (in five boxes and two large trunks) containing skins, "earths, salts, and minerals," and ethnographic materials, With them were a cage containing "four living Magpies," a prairie dog, and a sharp-tailed grouse. Amazingly, one magpie and the prairie dog survived the entire journey down the river to St. Louis, then to New Orleans and, by ship, to Baltimore. (The grouse made it as far as St. Louis.)

The materials brought back from the West belonged in principle to the government and the people of the United States. There was no national institution to receive them, however, and Jefferson (partly) appropriated them as his own and (mostly) undertook to place them where they would have scientific impact. The dried plant materials and the mineral and fossil specimens were deposited at the American Philosophical Society. Much of the ethnographic material was given to Charles Willson Peale to display in his Philadelphia Museum. In the subsequent checkered history of Peale's museums, most of the material was lost, but a few items found their way to the Boston Natural History Society and eventually to Harvard University. At the Peabody Museum of Archeology and Ethnology still at hand are "three Sioux raven belts, an otter bag, two basket weave whaler's hats from the Nootka or Makah of the Northwest, and a bear-claw necklace . . . a Mandan eagle-bone whistle and buffalo robe, a Sac and Fox tobacco pouch, a flute from the northern Plains Indians . . . a Crow cradle, and an Ojibwa knife."[12] The Harvard Museum of Comparative Zoology has the skin of Lewis's "black woodpecker."[13]

Lewis brought the expedition's collection of seeds to Philadelphia himself. They included pea, sunflower, parsnip, plum, honeysuckle, currant, serviceberry, flax, and tobacco seeds. Jefferson decided that they should be entrusted to two prominent Philadelphia plantsmen, Bernard McMahon, the commercial grower, and William Hamilton, the wealthy private collector and propagator. "Capt. Lewis has brought a considerable number of seeds of plants peculiar to the countries he has visited. I have recommended to him to confide principal shares of them to Mr. Hamilton of the Woodlands & yourself."[14]

Jefferson grew the Osage orange at Monticello. Because of the thorny bark of the young trees, it soon became popular for use in hedges. He also experimented with many of the potential crop plants collected by the expedition, including Pawnee corn and the Arikara corn, together with Arikara beans, Mandan tobacco, red and yellow currants, and gooseberries.[15] McMahon developed all of them commercially, but none proved as high-yielding as established eastern varieties.

In certain respects the expedition was a failure for Jefferson personally. The mastodon and the great-claw were not found, making it ever more likely that Jefferson was wrong on the subject of extinction. Another myth was put to rest, too, one that Jefferson had found only too believable. He had reported to Congress that "about 1000 miles up the Missouri, and not far from that river," there existed "a *salt mountain!* The existence of such a mountain might well be questioned, were it not for the testimony of several respectable and enterprising traders who have visited it, and who have exhibited several bushels of the salt to the curiosity of the people of St. Louis, where some of it still remains . . . This mountain is said to be 180 miles long, and 45 in width, composed of solid rock salt, without any trees or even shrubs on it."[16] There was no such mountain, and not many people, other than Jefferson, had ever thought there was.

Another failure, this one of far broader implications, was that no easily navigable route from the Missouri River to the Pacific had been found. "Easily navigable" is the key. Lewis and Clark did traverse a Northwest Passage, but it required serious portaging.

Perhaps the greatest failure, however, was that the expedition did not immediately capture the public's imagination. Political foes saw the whole thing as a waste of money, as another of Jefferson's quixotic quests for scientific knowledge. Potential supporters were left in the dark because of the failure of something quite modern—publicity. A lot of public money had been expended, but for too many years there was no public account, no book, about the team's adventures for voters to appreciate. Several of the team members had in fact kept extensive diaries. Lewis was to write up an account of the expedition at Jefferson's request, but he died unexpectedly (and mysteriously, from either murder or suicide). Benjamin Smith Barton and Nicholas Biddle took on the job of editing the expedition journals, but nothing appeared in print until 1814, eight years after the return of the Corps of Discovery—and the same year that the collected plants were written up in England by Frederick Pursh.[17] Charles Willson Peale exhibited the Indian artifacts, but the main mission of the expedition had been geographical discovery, not the habits of savages. Today the expedition is seen as an extraordinary, pivotal accomplishment in exploration, surveying and geography, natural history, and the establishment of national identity. At the time, it seemed to fizzle out.

PART SIX

Philosophical Issues

CHAPTER SEVENTEEN

"His Theories I Cannot Admire"

DURING JEFFERSON'S SECOND YEAR in Paris, his friend Francis Hopkinson wrote from Philadelphia about the remarkable case of a certain Dr. Moyse, a blind Scottish chemist and "a Philosopher by Profession." "He arrived I believe about a Year ago at Boston and has come from thence to this City, giving public Lectures in Natural philosophy all the way. He spent the beginning of this Winter at New York, where he became very popular and a great favourite of the Ladies in particular who crowded to his Lectures, and happy was she who [could] get him to dine or drink Tea at her House. Having gone thro' his Course there and reaped no small Honour and Profit, he is now performing with us."

Moyse's reception in Philadelphia was even more frenzied. "The Ladies are ready to break their Necks after him. They throng to the Hall at 5 o'Clock for places, altho' his Lecture does not begin till 7. He has been blind from his Infancy, has made Philosophy his Study and is well acquainted with the present admitted Systems, adding sometimes Theories of his own, which he does however, with rather too much Arrogance."[1]

The extent of the popular fashion for natural philosophy, which extended well into the next century, can be judged from the following poem: *An Address to the Late Dr. Moyse, by the Ladies of Edinburgh, in Consequence of a Course of Lectures Given by Him in That City, 1795*. It begins:

Dear Doctor, let it not transpire
How much your lectures we admire;
How at your eloquence we wonder,

When you explain the cause of thunder,
Of lightning and electricity
With so much plainness and simplicity;
The origin of rocks and mountains,
Of rain and hail, of frost and snow,
And all the winds and storms that blow.[2]

And so on, for forty-six lines. The science on which Moyse lectured was largely that of Isaac Newton. Every educated man's library contained a copy of Newton's *Principia Mathematica* in English translation or (more frequently) one of the many books being written for the lay reader. Jefferson was particularly enthusiastic about *Cours de Physique, Experimentale et Mathématique* by Pieter van Musschenbroek, the Leyden physicist who was one of the fathers of modern electricity. In fact, explaining Newton had became almost a cottage industry—hence the success of Dr. Moyse.

Newton, a devout if mystic Christian, opened the eyes of ordinary citizens to the vast range of possibilities of causal explanations of the material world and to the discovery of laws that controlled the motions of atoms. To many, there now seemed to be no room for miracles: no stopping the sun in its course. Galileo had been right in insisting that the relative movements of earth and sun were under physical control, and his tormentors were right when they realized that Copernicus's and Galileo's work relegated God to the role of putative creator; he no longer seemed the everyday controller of the universe. There was more than just a touch of blasphemy, however, in the epitaph that Alexander Pope suggested for Newton's grave: "Nature and Nature's laws lay hid in night: God said, 'Let Newton be!' and all was light."

The widespread popular interest in the exciting new natural philosophy and natural history had its downside. Newton himself was aware of the danger that his ideas might be seen to be supplanting God. Traditional beliefs were now open to be questioned by the new freedom of anyone to exercise his gift of reason; the authority of the church was challenged by science. And as science became a new and powerful source of authority it was bound to run headlong into the realms of existing authorities, particularly religious authorities. A conservative anti-intellectualism also ran through both European and American society. Resistance to change and to novelty therefore spread as discoveries mounted and as intellectual freedoms were exercised. Whether it was the quiet contrariness of Quakers and other Protestants, the profound philosophies of Descartes, Hume, and Newton, or the revolutionary ideas of Voltaire and Thomas Paine, there was ample reason to distrust them all.

The America of Thomas Jefferson and John Adams was a pragmatic place. By no means did every American intellectual share Jefferson's equal passion for both the useful and the more abstract aspects of natural philosophy. Attitudes changed in the quarter-century after the heady days of 1776, when philosophy and practicality had moved hand in hand. In 1800, Jefferson's intellectualism seemed out of joint with the times. For many, opinion had reverted to that expressed by a young man named William Scales (later a fellow of the Royal Society of London), who matriculated at Harvard in 1768, thirteen years after Adams. He was appalled at what he found there: "a seminary of sophistry, falsehood and folly . . . as soon as I applied myself to study, great Locke was delivered to me, to study whom I found to be a miserable destroyer of the understanding; after that renowned Sir Isaac Newton came before me for examination, and I found him a great fabricator of falsehoods, and a destroyer of the word of God."[3]

To put into context Jefferson's interest in the philosophical side of science and the opposition it encountered, it is instructive to consider the views of his friend, ally, and rival John Adams. Adams was eight years older than Jefferson and came from a totally different world, geographically, economically, and intellectually. Like Jefferson, he thought deeply about things, especially law, literature, and philosophy, as his early diaries show. But he was an entirely different kind of intellectual.

Adams could be opinionated about the more philosophical parts of science, and his words provide an important counterpoint to Jefferson's enthusiasms and perhaps give a more accurate index of the temper of the whole country. His most famous statement on the subject was contained in a letter to his wife, Abigail, in 1780: "I must study Politicks and War that my sons may have liberty to study Mathematicks and Philosophy. My sons ought to study Mathematicks and Philosophy, Geography, natural History, Naval Architecture, navigation, Commerce and Agriculture, in order to give their Children a right to study Painting, Poetry, Musick, Architecture, Statuary, Tapestry and Porcelaine."[4] Typically, the message was ambivalent; he revealed regret for what he had nobly sacrificed and simultaneously conveyed a sense that the world was tending toward a loss of seriousness.

As a student at Harvard College, Adams was fired by Winthrop's lectures on natural philosophy and "began to feel a desire to equal" his contemporaries in science and literature. "In the Sciences especially Mathematicks, I soon surpassed them, mainly because, intending to go into the Pulpit, they thought Divinity and the Classics of more Importance."[5] In later years, "my Smattering of Mathematics enabled me afterwards at Auteuil in France to go, with my eldest Son, through a course of Geometry, Algebra and several Branches of the Sciencs, with a degree of pleasure."[6] In a letter to his lifelong friend Dr. Benjamin Waterhouse, Adams admitted that "I then attempted a sublime flight and endeavoured to give him some Idea of the Differential Method of Calculation of the Marquis de L'Hospital, and the Method of Fluxions and infinite Series of Sir Isaac Newton. But alas it is thirty years since I thought of Mathematicks, and I found I had lost the little I once knew."[7]

At age twenty-one he wrote in his Diary, "I long to prosecute the mathematical and philosophical sciences."[8] But Adams had a clear idea of where he was going and knew there was not time enough to study everything. "I have so many Irons in the Fire, that every one burns.—I have common, civil, natural Law, Poetry, Oratory, in Greek, latin, French, English to study, so that when I sit down to read or think, so many subjects rush into my mind that I know not which to chuse."[9]

Adams would no doubt have had the support of most Americans for his view that "Order, Method, System, Connection, Plan, or whatever you call it, is the greatest Proof of Genius, next to Invention of new Wheels, Characters, Experiments, Rules, Laws, which is perhaps the first and greatest."[10] For Adams, science was useless unless it had immediate practical ends. "The primary Endeavour . . . should be to distinguish between

Useful and unuseful, to pursue the former with unwearied Industry, and to neglect with much Contempt all the rest."[11]

In a typically dyspeptic letter to his friend Francis Adrian van der Kemp (a displaced Dutch intellectual and an enthusiast for "unuseful" facts, if ever there was one), Adams wrote that "nature itself is all arcanum; and I believe it will remain so. It was not intended that men with their strong passions and weak principles should know much. Without a more decisive and magisterial discernment, much knowledge would make them too enterprizing and impudent."[12]

Nor did the young Adams have much respect for moral philosophers: "Many Pens are employed, much Time spent and much Mischeif and Malevolence occasioned, by Divines about Predestination, [the] Original of Evil, and other abstruse subjects, that having been to no good Purpose under learned Examination so many Centuries may by this Time be well enough concluded unfathomable by the human Line." Instead, he wrote, "let the few who have been distinguished by greater intellectual Abilities than Mankind in general, consider, that Nature intended them for Leaders of Industry. Let them be cautious of certain Airs of Wisdom and superiority by which some Gentlemen of real sense and Learning, and Public spirit, giving offence to the common People, have in some Measure defeated their own benevolent Intentions. Let them not be too sparing of their Application or Expence, lest failing of visible Profit and success they expose themselves to Ridicule and rational Husbandry itself to Disgrace among the fair Progeny of Science, and the numerous functions of Fame."[13]

This last passage was written before he and Jefferson had been thrown together as colleagues at Philadelphia, but it could easily have been directed at Jefferson, whom Adams considered not at all "cautious of certain airs of Wisdom" and certainly not sparing when it came to "Expence." Jefferson's chief sin, in Adams's eyes was perhaps that, unlike Adams, he had the capacity to be devoted to natural philosophy while at the same time applying his knowledge to useful things.

Just as Jefferson found everything about the mastodon fascinating, from its identity and habits to the possibility that it still lived in toward the

Great Stoney Mountains, Adams was repelled. "Speculations about Mammoths . . . These are all pitiful Bagatells, when the Morals and Liberties of the nation are at hazard as in my Confusion I believe them to be at this moment. And the atheism of your Buffon and the despicable Philosophy of Mammoths . . . have made them so."[14]

As for Buffon, "I delight in Buffon's facts and his manner of relating them, when he is correct: but his Theories I cannot Admire."[15] Elsewhere, he wrote with telltale impatience and irascibility, "I care not a farthing about all the Big Bones in Europe or America. Nothing of this kind is to my taste."[16]

In 1774, however, Adams visited the natural history collection "made by a Mr. Arnold an Englishman of Birds, Insects, especially Butterlifes . . . This cabinet was afterwards sold to Governor Tryon of New York and then sent to London. In 1778, I went to France, where I saw many cabinets, and some of more curiosity and magnificence than Use: but they all served to impress upon my mind, the Utility of some Establishment in America for collecting Specimens of the Works of Nature curious to us."[17] Jefferson was already a patron of such a project: Charles Willson Peale's museum.

Neither Adams nor Jefferson, nor most of their contemporaries, could find a satisfactory explanation for meteorites, which became a much-discussed subject in America after 1807, when the Weston meteorite crashed through the roof of a house in Connecticut. One theory was that these "stones from the sky," as Benjamin Silliman at Yale called them, had spun off from the moon. Adams's objection was scientific: he wrote to van der Kemp that the origin of meteorites from a "Vulcano in the Moon, is almost as absurd as Buffon's Earth dashed off from the sun in the form of an ocean of melted glowing glass." Apart from disdain for Buffon's theories, Adams's objection reflected what he had learned on Newtonian mechanics at Harvard. "All the Powers of Matter in the Universe could not give a whirling motion, a rotation round the Axis a revolution round the Sun, nor any circular or curved motion to this glass or any part of it. Force or Power impressed upon matter impells it in right lines and no other." Significantly, Adams ended with the cutting words: "You should correspond with Jefferson upon these points."[18]

Jefferson, also a skeptic on the subject of meteorites having an extra-terrestrial origin, wrote cautiously to a friend who had collected one: "[Its] descent from the atmosphere presents so much difficulty as to require careful examination . . . We are certainly not to deny whatever we cannot account for. A thousand phenomena present themselves daily which we cannot explain, but where facts are suggested, bearing no analogy with the laws of nature as yet known to us, their verity needs proofs proportioned to their difficulty . . . Is it easier to explain how it got into the clouds from which it is supposed to have fallen?"[19]

When Jefferson and Adams's correspondence resumed in later life, Jefferson got into a discussion with Adams about one of his favorite subjects, the origins of the various American Indian peoples. Did they come from Asia, as their physical features indicated, or did their relatives in Asia come originally from America? That was the sort of question that Adams abhorred, as he witheringly demonstrated: "Whether Serpents Teeth were sown here and sprung up Men; whether Men and Women dropped from the Clouds upon this Atlantic Island; whether the Almighty created them here, or whether they immigrated from Europe, are questions of no moment to the present or future happiness of Man. Neither Agriculture, Commerce Manufactures, Fisheries . . . will be promoted, or any Evil be averted, by any discoveries that can be made in answer to those questions."[20]

Perhaps the most dramatic example of the gap between Adams's and Jefferson's views of science concerned miracles. For the ultra-rationalist Jefferson the concept of miracles was the remnant of a primitive superstition, perpetuated by a controlling clergy. Matter was subject to inviolable laws, not a deity's whim. As he had written to Peter Carr, "You are Astronomer enough to know how contrary it is to the law of nature that a body revolving on it's axis, as the earth does, should have stopped, should not by that sudden stoppage have prostrated animals, trees, buildings, and should after a certain time have resumed it's revolution, and that without a second general prostration. Is this arrest of the earth's motion, or the evidence which affirms it, most within the law of probabilities?"[21]

Adams, on the other hand, while he was keenly interested in astronomy, believed implicitly in the historical accuracy of biblical narratives. For him, miracles were real. As a young man, he wrote in his diary: "The great and Almighty author of nature, who at first established those rules which regulate the world, can as easily suspend those laws whenever his providence sees sufficient reason for such suspension."[22] Belief did not, however, deflect his strong interest in science. Soon after graduating from Harvard, for example, he became engaged in a controversy that popped up over Franklin's lightning conductors between Professor Winthrop and a Boston preacher, the Reverend Thomas Prince. Prince had suggested that a recent earthquake was due to the increasing installation of "iron points," which, attracting lightning, caused electricity to be stored in the ground, eventually causing the earth to burst asunder. "This Invention of iron Points to prevent the Danger of Thunder has met with all that Opposition from the Superstition, Affectation of Piety, and Jealously of New Inventions, that Innoculation to prevent the Danger of the Small Pox, and all other useful discoveries, have met with in all Ages of the World."[23]

As a farmer, Adams was interested in the uses of seaweed. He devoted a whole page of his youthful diary to notes on the different kinds of kelp, how kelp grew, and the process of burning it for use as fertilizer, noting just as Jefferson would have that "20 Tons of the Weeds will produce about one Ton of the ashes."[24]

The range of Adams's scientific interests shows up in his correspondence with his friend Benjamin Waterhouse. In France, Adams had followed modern discussions about animal respiration and the composition of air. When Waterhouse later published a book entitled *Animal Life,* Adams wrote excitedly to him. "It has long appeared to me astonishing, that it should be impossible to discover, what it is, which the Air conveys into our Lungs and leaves behind it, in the Body when we breathe." Waterhouse subscribed to a vitalist view of life. Adams wrote skeptically, "Pray where is the Evidence of the Existence of a subtle Electric fluid which pervades the Universe . . . where is your Authority for saying that such an Electric fluid is the Cause of Life.? Why may it not as well be Magnetism? Or Steam, or Nitre? Or fixed air?"[25]

During his time in France he, along with everyone else, became interested in Frederick Anton Mesmer. He wrote to Waterhouse: "All Paris, and indeed all Europe, is at present amused with a kind of physical new Light of Witchcraft, called Animal magnetism . . . [Mesmer] pretends that his Art is an Universal Cure, and wholly supersedes the Practice of Physick and consequently your Professorship, so that you will not, I hope, become his Disciple." Adams recognized that "the Professors of this Art have acquired sometimes a surprising ascendency over the Imaginations of their Patients, so as to throw them into violent Convulsions, only by a few odd Gestures . . . I think you Physicians ought to study and teach us some Method of managing and controulling it."[26]

Perhaps the most intriguing of Adams's scientific fascinations concerned the "generation of Shell fish." He corresponded at length with both Waterhouse and van der Kemp about the reproduction of sea life, including both shellfish and jellyfish. On his voyage home from France he had the crew net some of the creatures for him. He described them in detail, including the stings of the jellyfish, which he wondered might use some kind of electricity. Having collected jellyfish at different life stages, he could outline some of their growth and development. He also speculated on the method by which hard shells might be formed and grow and, more generally, on the means by which marine invertebrates—billions upon billions of them—came to be so prolifically distributed in the oceans.[27]

This interest in sea life is surprising in the man who commented, as a youth, that he had no time for those natural philosophers who "have employed the Advantages of great Genius, Learning, Leisure, and Expense, in examining and displaying before the World, the formation of Shells, and Pebbles, and Insects, in which Mankind are no more interested, than they would be in a laborious Disquisition into or sage Conjectures about the Number of sands in the Moon or of Particles in the solar system."[28]

Jefferson had a greater interest in astronomy than Adams did. After all, Jefferson owned various telescopes for observing the heavens, and his letters show a fascination with William Herschel's discovery of the planet Uranus. But when he and Adams made a trip to England together in 1786, it was Adams, not Jefferson, who went out of his way to meet Herschel and to inspect his impressive new telescope.

Digressions and odd enthusiasms notwithstanding, in the realm of science and philosophy, Adams tended to remain the practical man and Jefferson the intellectual. They were at opposite poles of American attitudes, and by the end of the eighteenth century such differences came to be expressed politically.

CHAPTER EIGHTEEN

Philosophers Unwelcome

WHEN THOMAS PAINE, A populist hero during revolutionary times, intended to return to the United States from France after twenty-five years away, an anonymous writer in *Port Folio* fulminated, "What! Invite to the United States that lying, drunken, brutal infidel."[1] There, in a nutshell, was a view of contemporary continental philosophers typical of those who saw Jefferson and his allies as embracing every foreign, radical, and godless threat.

For Jefferson's Federalist opponents, the more abstract elements of natural philosophy were not only a waste of time, they were dangerous. The origins of their suspicion lay partly in religion and partly in political philosophy. In their view, the most immediate threat, greater even than a call to the literal barricades, was the insidious, sometimes open tension between the new philosophies, with their scientific explanations of causes in the material world, and religion. Modern philosophies, especially those coming out of France, were seen as leading to a denial of basic tenets of Christianity—as indeed, in Jefferson's case, they did—and as threatening the stability of the young nation.

One author lambasted Jefferson and others like him on precisely these grounds: "Wretched, indeed, is our country, if she is to be enlightened by . . . philosophers whose industry is equalled by nothing but their vanity; whose pursuits are impeded by no danger nor difficulty . . . who think nothing too great for them to grasp, and nothing too minute to be observed: they dig into the bowels of the earth, and climb the loftiest mountains; they traverse the ocean, and explore the region of air; they search the written records of antiquity, and the traditions of savages; they built up theories of shells and bones and straws. And for what? Is it to render more stable the uncertain condition of man? . . . No; it is to banish civilization from the earth . . . to extinguish the only light by which the Christian hopes to cheer the gloomy hour of death . . . to degrade us from the rank of angels . . . that we may complete the catalogue of brutes."[2]

A culture war is evident here. A rich vein of anti-intellectualism runs though all of American life. But there was also an understandable alarm at what was happening in Europe and the way that it had been fomented. For many, France, which had been America's earliest and most valuable ally, had become, at the end of the eighteenth century, the enemy. From a pulpit in New Haven, Connecticut, President Timothy Dwight of Yale College, thundered in his July 4 sermon of 1798: "About the year 1728, Voltaire, so celebrated for his wit and brilliancy, and not less distinguished for his hatred of Christianity and his abandonment of principle, formed a systematical design to destroy Christianity, and to introduce in its stead a general diffusion of irreligion and atheism."[3] The result, Dwight stated, was that "no personal, or national interest of man has been uninvaded; no impious sentiment, or action, against God has been spared . . . Justice, truth, kindness, piety, and moral obligation universally have been, not merely trodden under foot; this might have resulted from vehemence and passion; but ridiculed, spurned, and insulted as the childish bugbears of driveling idiocy . . . Nor has any art, violence, or means, been unemployed to accomplish these evils."[4]

Dwight's sermons gave voice to the great fear that the French Revolution, stage-managed by men who had radical ideas but no practical experience in government, might be imported to America. The political differences over this in America were profound. Jefferson always erred in favor of revolution, almost no matter the cost in blood, if it were the will of the majority of the people. Calmer heads hankered after authority and experience (which had a tendency to mean the British way of things).[5] Many thought that perhaps it was a mistake to put too much power in the hands of "the people."

As the "Marseillaise" was being sung in France, it was all too easy to see continental-style philosophers as the enemies of Christianity. Dwight developed this theme in a sermon in 1801 against these unbelievers, heavily italicized for emphasis. "The Infidels, here referred to, are plainly *philosophists; the authors of vain and deceitful philosophy; of science falsely so called; always full of vanity in their discourses . . . and alluring others . . . to follow them; promising them liberty, as their reward, and yet being themselves, and making their disciples, the lowest and most wretched of all slaves, the slaves of corruption. Philosophical pride, and the love of sinning in security and peace, are, therefore, the two great causes of Infidelity.*"[6]

Religious freedom was, for Jefferson, a fundamental right. He drafted the Virginia Statute for Religious Freedom (submitted in 1779, enacted in 1786), the principles of which can be seen as foreshadowing the Declaration of Independence. He considered it one of most important contributions to American democracy.[7] Like Franklin, Jefferson was a deist, not a conventional Christian, but he had strong convictions about a Creator.[8] In the letter to Adams of April 11, 1823, he wrote, "I hold (without appeal to revelation) that when we take a view of the Universe, in it's parts general or particular, it is impossible for the human mind not to percieve and feel a conviction of design, consummate skill, and indefinite power in every atom of it's composition. The movements of the heavenly bodies, so exactly held in their course by the balance of centrifugal and centripetal forces, the structure of our *earth* itself, with it's distribution of lands, waters and atmosphere, animal and vegetable bodies, examined in all their minutest particles, insects mere atoms of life, yet as perfectly organised as man or mammoth, the mineral substances, their generation and uses, it is impossible, I say, for the human mind not to believe that there is, in all this, design, cause and effect, up to an ultimate cause, a fabricator of all things from matter and motion, their preserver and regulator while permitted to exist in their present forms, and their regenerator into new and other forms."[9]

Jefferson held to a rather nonaggressive form of deism. He largely kept his religious ideas to himself, developing a set of rational beliefs that largely retained the moral teachings of Jesus but rejected as unnecessary all the miracles, the virgin birth, and perhaps even—heresy of heresies—the Resurrection itself. As Thomas Paine, never subtle, wrote in 1805, "When the divine gift of reason begins to expand itself in the mind and calls man to reflection, he then reads and contemplates God and his works, not in the books pretending to revelation . . . The little and paltry, often obscene, tales of the bible sink into wretchedness when put in comparison with this mighty work. The deist needs none of those tricks and shows called miracles to confirm his father, for what can be a greater miracle than creation itself, and his own existence."[10]

Soon after leaving William and Mary College, Jefferson copied into his Literary Commonplace Book long passages from Henry St. John Bolingbroke's philosophical essays, including the following trenchant comment on the central doctrine of conventional Christianity: "God sent his only begotten son, who had not offended him, to be sacrificed by men, who had offended him, that he might expiate their sins and satisfy his own anger, surely our ideas of moral attributes will lead us to think that god would have been satisfied, more agreeably to his mercy and goodness, without expiation, upon the repentance of the offenders, and more agreeably to his justice with any other expiation than this."[11]

In 1803, Jefferson quietly prepared a document entitled "Syllabus of an Estimate of the Merit of the Doctrines of Jesus, Compared with Those of Others." His aim was to distill the essence of Jesus's teachings, free of miracles and mysteries. Unfortunately, he lent a copy of his manuscript to Adams's friend van der Kemp, who allowed it to be published in Boston. When revealed to be the author, Jefferson, never uncontroversial, became even more suspect in the eyes of ardent Christians, whose fundamentalism and biblical literalism grew year by year. His friendship with the exiled English preacher and chemist Joseph Priestley did nothing to change that opinion among his enemies. Jefferson later created his own version of the New Testament Gospels, editing them down to a basic set of passages that he considered rational and worthy of following, and omitting what he and other deists considered unnecessary and insupportable superstitions and fables.[12]

History shows over and over again that what seems logical and sensible to one person may seem arcane, contrived, irreligious, or even treasonous to another. One person's passion can easily be ridiculed, satirized, and mocked by another. Any strength can be turned by perhaps perverse logic into a weakness. Accordingly, both Jefferson's deism and his science became wonderfully useful sticks to beat him with.

Jefferson's religion was intolerable to church elders, especially in Federalist New England and New York. One of his most outspoken critics was the Reverend Clement C. Moore—he of the gentle poem "'Twas the Night before Christmas."[13] There was nothing childlike about the excoriating attack on *Notes on the State of Virginia* that Moore published in 1804.[14] His intention was to highlight those passages that "tend to the subversion of religion; and to examine whether, from brilliancy of invention, acuteness of investigation, or cogency of argument, they are entitled to the name of any other than modern French philosophy"—"modern French philosophy" being a term of utmost opprobrium. Although Jefferson had been careful to qualify his statements on the origins of mountains and to set them within the accepted biblical account of creation, Moore read him as an out-and-out atheist. Seizing on Jefferson's statement that the earth was "created in time," he accused him of offering a "theory of the earth contrary to the scripture account of the creation," one that "denies the possibility of an universal deluge" and "considers the Bible history no better than ordinary tradition."

"Whenever modern philosophers talk about mountains," Moore observed, "something impious is likely to be at hand." Jefferson's geology "seems to posses every qualification which the heart of a modern philosopher could desire; it is bold, plausible, and contrary to Scripture." *Notes,* he said, "extols Voltaire and the French Encyclopedists, the imps who have inspired all the wickedness with which the world has of late years been infested."

Moore did not forget Jefferson's denigration of blacks and his unfortunate reference to the "orang-outang." If his theory "does not exalt the orang-outang to the station of a rational being, it debases negroes to an order of creatures lower than those who have a fairer skin and thinner lips." Moore quoted the explorer Mungo Park to the effect that blacks "appear not only susceptible of the purest love, but many of them possess hearts so generous and compassionate, that civilized nations might profit by their example. Although they have not 'been liberally educated,' nor 'lived in countries where the arts and sciences are cultivated to any considerable degree,' they have carried their arts to a much higher degree of excellence."

Even Jefferson's familiarity with the lowly creatures of America could be used against him. The journalist Luther Martin, the son-in-law of the Colonel Cresap whom Jefferson, in later editions of *Notes,* accused of murdering Chief Logan's family, took to the newspapers a long letter in which he mocked Jefferson as a "philosopher . . . employed in weighing the rats and mice of the two worlds to prove that those of the *new* are not exceeded by those of the *old.*" As for the American Indians whose side Jefferson took in the Cresap affair, Martin complained that Jefferson had examined minutely "*every part* of their frame." He followed this slightly risqué hint that Jefferson had measured the genitals of Indians and found them no weaker or smaller by quoting from a pirated French edition of *Notes* in order to give added piquancy and a sense of foreign flummery: "ses organs de la generation ne sont pas plus foibles ou plus petits."[15]

The Louisiana Purchase and the expedition of Lewis and Clark were similarly criticized, both as a whim of Jefferson's and as a squandering of money. (Adams was one of the few Federalists to support the Louisiana Purchase, arguing that it gave America control of the Mississippi navigation.)[16] Jefferson had not helped himself by hoping that a living mastodon or great-claw would be found, or the "salt mountain."

Turnabout is fair play, or, in this case, unfair play. The Democrat-Republicans did not hesitate to circulate charges about their opposition. Even so, it is hard to imagine who thought up the idea of putting out the story that the sober John Adams had sent the brother of Charles Pinckney to England to procure four mistresses—two for himself and two for Adams. Adams employed the very best defense—humor: "I do declare if this be true, General Pinckney has kept them all for himself and cheated me out of my two."[17]

For Jefferson, science, philosophy, and education were the keys to social progress. However, progress was threatening to many, if not most, Federalists. They were not comfortable with the idea that what had been so hard won and seemed the very fiber of the American people was susceptible to change. For who was to say what was forward progress and what was backward? Readers of history like Adams could easily find precedent for their view that progress did not always lead to the greater good but often led to moral and civil decay, that increased prosperity produced political and mercantile corruption. The evidence of the French Revolution showed only too clearly what happened when power fell into the hands of populist zealots.

The best that Adams could say about "progress" was what he once wrote to Benjamin Waterhouse: "My humble opinion is that Sciences and Arts have vastly and immensely ameliorated the condition of Man, and even improved his Morals. The progress, however, has been awfully slow."[18] But Adams's political allies had a different view. As Jefferson recalled, "One of the great questions, you know, on which our parties took different sides, was on the improvability of the human mind in science, in ethics, in government, etc. Those who advocated reform of institutions, pari passu, with the progress of science maintained that no definite limits could be assigned to this progress. The enemies of reform, on the other hand, denied improvement, and advocated steady adherence to the principles, practices and institutions of our fathers, which they represented as the consummation of wisdom . . . You predict that [the freedom of enquiry] will produce nothing worthy of transmission to posterity, than the principles, institutions, and systems of education received from their ancestors. I do not consider this as your deliberate opinion. You possess yourself, too much science, not to see how much is still ahead of you, unexplained and unexplored."[19]

Adams's reply included his famous admonition: "Checks and Ballances, Jefferson, however you and your Party may have ridiculed them, are our only Security, for the progress of Mind, as well as the Security of Body."[20]

The event that had brought to a boil so much of the opposition to Jefferson's scientific and political philosophy was the presidential and vice presidential election of 1800. Americans had discovered a new way to become divided against each other—party politics, the very thing that George Washington had feared. Jefferson and Aaron Burr stood for the Democrat-Republicans against Adams and Charles Pinckney for the Federalists. Jefferson and Burr tied with the most votes, and Jefferson then prevailed in the House of Representatives. Suddenly rival philosophies had ceased to be something abstract, pursued in private by a few; they had become a major public issue.

Federalists feared that the election of Jefferson would mean a "War abroad [with Britain, presumably], and the introduction of French influence; a change in the constitution—a general overthrow of our internal policy, and discord at home."[21] The level of vituperation in the 1799 campaign and the "Revolution of 1800" makes modern party politics seem tame. A low point came when an unknown journalist proclaimed that "Thomas Jefferson . . . to make the best of him, was nothing but a mean-spirited, low-lived fellow, the son of a half-breed Indian squaw, sired by a Virginia mulatto father" and raised "wholly on hoe-cake . . . bacon, and hominy, with an occasional change of fricasseed bullfrog, for which abominable reptiles he had acquired a taste during his residence among the French in Paris."[22]

Timothy Dwight took on the philosophers whom Jefferson favored. "For what end," he asked," shall we be connected with men, of whom this is the character and conduct? . . . Is it, that we may see our wives and daughters the victims of legal prostitution; soberly dishonoured; speciously polluted; the outcasts of delicacy and virtue?"[23]

Having—with Jefferson in the lead—gotten rid of one set of establishment hierarchies (church and crown), Americans now found themselves unwilling to embrace a new aristocracy, that of a scientific and intellectual (philosophical) elite. In a sense, Jefferson and his colleagues in science had become the victims of their own success. It was many years before the situation could be reversed.

Jefferson succeeded in his quest for election as the country's third president but found that his years in office (1801–1809) afforded him little time or opportunity for scientific pursuits, especially natural philosophy. He kept up his climatological measurements and, as earlier noted, continued almost compulsively to make detailed lists of things like the seasonal appearance of different vegetables in the capital's market. In connection with the Lewis and Clark Expedition he became entangled for a considerable time in a version of a puzzle being tackled (unsuccessfully) by so many across Europe: finding a way reliably and easily to determine longitude without the use of an accurate chronometer. Here his mathematical skills and astronomical knowledge failed him. He tinkered with improvements to pedometers and followed closely the latest developments in steam navigation. He also came up with the idea of maintaining a fleet of naval vessels in reserve in special dry docks.[24] Doing so would have reduced the need to keep a large active fleet but was an entirely impractical notion for ships whose wood would dry out and whose seams would open.

Jefferson seems always to have had a low opinion of medical science. There was a family doctor in Charlottesville, Dr. William Wardlaw, but Jefferson liked to be in charge of treatments for his family and his slaves. In 1801 he decided to adopt the new, very successful, practice of vaccination against smallpox developed by Dr. Edward Jenner in England and pioneered in America by Dr. Benjamin Waterhouse at Harvard.[25] He had his family and his slaves vaccinated and encouraged all the members of his scientific circle to promote the technique in their home states. He even persuaded a visiting Miami chief, Little Turtle, to adopt it for himself and his people and, shortly afterward, a delegation of Delaware and Shawnee as well.[26] As previously noted, Lewis and Clark took the wherewithal for vaccinations on their expedition.

Jefferson was philosophically committed to having smaller rather than bigger national government, to reducing the national debt, and to abolishing federal taxes. In line with this thinking, he opposed Alexander Hamilton's scheme for a national bank, and he did not, despite his personal interest in science, encourage direct federal involvement in the sciences. The American Philosophical Society in Philadelphia and its sister organizations in New York and Boston were the leading promoters of science. There was no other focused source for scientific advice to the government until the founding of a National Academy of Sciences in 1863. It was to his colleagues at the American Philosophical Society that Jefferson turned for help in adjudicating claims for a new process for desalination of water and for coming up with a solution to the Hessian fly problem.[27]

Right from the beginning, a few federal responsibilities had been largely agreed on. As early as 1786, for example, it had been obvious that standardization of weights and measures was essential both for interstate commerce and for international trade, for which a national currency was needed. Secretary of State Jefferson laid out his "Plan for Establishing Uniformity in the Coinage, Weights, and Measures of the United States" in 1790, and Congress created the United States Mint in 1792. Despite Jefferson's inclinations toward decentralization, two new national organizations were founded during his administration. The United States Military Academy (West Point, 1802) was established as a place to study the sciences of war, and the Coast and Geodetic Survey (1807) was established to apply science to practical ends—in this case, navigation and the promotion of commerce.[28]

Although the Louisiana Purchase, which Jefferson stage-managed in 1803, doubled the size of the United States and gave it control over the Mississippi and Missouri Rivers, the main arteries of the continent, and New Orleans, the portal to the West, the addition of the western lands brought a new dilemma with respect to American Indians and African Americans. Now even more native peoples would inevitably be displaced, and the question of what role slavery would play in the new territories reared its head. Jefferson did not change his mind now that he was president. As he had written in *Notes on the State of Virginia,* he believed that Indians should only be assimilated or banished to Canada (or farther) as the settlers poured into the West to set up their farms and towns. There would be no haven set aside for them even in those seemingly limitless lands. Nor would land be allotted to emancipated people. Such an experiment would be too dangerous to try. Jefferson's view of the United States was one of (white) racial homogeneity. In March 1807 he did, however, sign the law banning the slave trade (specifically, the importation of slaves from abroad).

During his second term in office Jefferson became more and more embattled politically and personally and ever more frustrated that affairs of state and the insidious drip of petty politics left less and less room for serious philosophy or science. One source of stimulation for his intellect was dinner parties at the White House where he could associate with like-minded colleagues and enjoy fine food and French wines. John Quincy Adams, son of John Adams and the future sixth president, though no lover of Jefferson's politics, dined several times with Jefferson, and his diary shows that conversation at the table tended often to range far and wide. "You never can be an hour in this man's company without something of the marvelous."[29] Adams claimed that Jefferson was not above dramatic exaggerations in his storytelling. Jefferson had, for instance, recalled that the temperature in Paris never rose above zero degrees Fahrenheit for a stretch of six weeks. Jefferson's credulity was not improved, Adams wrote, by his insistence that "zero . . . is fifty degrees below the freezing point."[30] This mistake seems one that Jefferson could not have made except possibly to tease his guests.

At the dinner on November 3, 1807, Jefferson and Dr. Samuel Latham Mitchell (congressman and scientist) led the table in discussions of "Epicurean philosophy, Fulton's steamboat and underwater torpedo, chemistry, geography, natural philosophy, oils, grasses, beasts, birds, petrifactions and incrustations, Pike, Humboldt, Lewis and Barlow . . . Mr. Jefferson said that he had always been keen on agriculture, and knew nothing about it, but that the person who united with other science the greatest agricultural knowledge of any man he knew was Mr. Madison . . . On the whole it was one of the *agreeable* dinners I have had at Mr. Jefferson's."[31]

At these dinners, Jefferson insisted on dressing informally, which created something of a scandal in the diplomatic community. At one point he expressed his displeasure that the French minister had come to dine with "a profusion of gold lace on his clothes. He [Jefferson] says they must get him down to a plain frock coat."[32] Given the trivialities of life in Washington, party factionalism, and America's difficult relations with Britain and France, to say nothing of the problem of dealing with the Barbary pirates' interference with shipping, it is not surprising that in 1809, when his second term was over, Jefferson escaped with relief to the relatively orderly chaos of rebuilding Monticello and the last great task of creating the University of Virginia.

Despite his insistence on the value of education, Jefferson was not in a position to mandate a single standard for the whole country, nor would he have been philosophically in favor of federal intervention. But it is clear what he thought was important. In Virginia, as early as 1779, Jefferson had promoted universal education through the establishment of schools in every county; at them "shall be taught reading, writing, and common arithmetick, and the books which shall be used therein for instructing the children to read shall be such as will at the same time make them acquainted with Græcian, Roman, English, and American history. At these schools all the free children, male and female, resident within the respective hundred, shall be intitled to receive tuition gratis, for the term of three years, and as much longer, at their private expence, as their parents, guardians or friends, shall think proper."[33] He envisaged three successive levels of free education: elementary school, grammar school, and college, with "sciences" to be taught at the college level, including at the College of William and Mary (where a three-tier scheme was already in place). The scheme was far too ambitious and costly, but a version of it was finally passed by the Virginia General Assembly in 1796.

In a long 1779 proposal for redefining the role and curriculum of the College of William and Mary, Jefferson had laid out a curriculum for the college that looked very much like the scheme by which his library was organized.[34] In the sciences, he prescribed "Mathematics (Pure): Arithmetic. Geometry Mechanic." In "Mathematics (Mixed)" were "Optics. Acoustics. Astronomy." Then came "Anatomy and Medecine." Under Natural Philosophy he listed "Chymistry. Statics, Hydrostatics. Pneumatics, and Agriculture." Under Natural History fell "Animals—Zoology. Vegetables—Botany. Minerals—Mineralogy." All these subjects were to be taught by eight professors, "to wit, one of moral philosophy, the laws of nature and of nations, and of the fine arts; one of law and police; one of history, civil and ecclesiastical; one of mathematics; one of anatomy and medicine; one of natural philosophy and natural history; one of the ancient languages, oriental and northern; and one of modern languages."

Starting in 1813, Jefferson set forth his ideas for a new University of Virginia, extending the ideas he had laid out for the College of William and Mary: "It is the duty" of our country "to provide that every citizen in it should receive an education proportioned to the conditions and pursuits of his life."[35] His model was still highly stratified. There should be elementary schools from which two streams of students would emerge. Those "destined for labor" would go to work or to apprenticeships, and "their companions, destined to the pursuits of science, will proceed to the college." The college would exist in two levels, general and professional. The second "learned class" would be made up of those preparing for professions and those from wealthy families who "may aspire to share in conducting the affairs of the nation, or to live with usefulness and respect in the private ranks of life."

For curriculum and professors, Jefferson stuck to the old Baconian classification of the branches of knowledge. Jefferson thought that his "learned" class would require training in three areas: languages (from ancient language to "belles letters and rhetoric" and "history"), mathematics (in which department he included physics, chemistry, natural history, anatomy, and medicine), and philosophy (ethics, laws of nature and of nations, government, political economy). All subjects were to be taught by four professors.[36]

Of longer-lasting significance for the new university than his plans for classes was Jefferson's design for the buildings: an "academical village" with "pavilions" for professors and students (and their slave servants). To the visitor today, a particularly striking feature is the perfect balance of the structures. The design of the magnificent central Rotunda, with its yet more superb central stairway, was based on the Pantheon in Rome, but was scaled down to fit with the march of the pavillions along the lawn. It stands now as the crown jewel of this meeting of art and science.

CHAPTER NINETEEN

Transcendental Truths

JEFFERSON WAS CONVINCED OF the power of natural philosophy, through reason, experiment, and observation, to produce new necessary truths that transcended time, place, and fashion. The kind of natural philosophy that he loved was at its height in the mid to late eighteenth century. After that, it was gradually transformed by division into the separate subjects of modern science and partially eclipsed by technology, at which point it lost connection to philosophy itself. At its peak, natural philosophy, and particularly Newtonian mechanics, was used to illuminate a whole range of nonscientific subjects. The preservation of consistency and stability through the operation of offsetting forces—as in the case of the regular orbits of the planets around the sun—became a particularly powerful political metaphor.[1]

In 1728, for example, John Theophilus Desaguliers, a Frenchman who had been Newton's student and who devoted a good deal of his life to explaining and promoting Newton's works, wrote a long allegorical poem using the harmony of the cosmos as a metaphor for the best model of government. "The *limited Monarchy,* whereby our Liberties, Rights, and Privileges as so well secured to us . . . seems to be a lively Image of our System; and the Happiness that we enjoy under *His* present MAJESTY'S Government, makes us sensible, that ATTRACTION is now as universal in the Political, as the Philosophical World." ("Attraction" was an early term for gravity.)[2]

It is fitting, however, that a chapter of this book on Jefferson's science in the context of a world of ever-changing ideas should deal with one supremely prominent case in which Jefferson did not think as a scientist.

Jefferson's devotion to the principles of natural philosophy remained complete through his lifetime. In the heady days of 1776, philosophy and practicality had seemed to work hand in hand. A question has therefore been raised: Is there any Newtonian imagery or other influence of contemporary natural philosophy in Jefferson's greatest piece of writing, the Declaration of Independence? Given the wide influence of natural philosophy and Jefferson's scientific predilections, this is not a trivial question, and the answer helps put eighteenth-century science into context.

In the Declaration of Independence, Jefferson's immortal words transcended time and place. "When in the Course of human events, it becomes necessary for one people to dissolve the political bands which have connected them with another, and to assume among the powers of the earth, the separate and equal station to which the Laws of Nature and of Nature's God entitle them, a decent respect to the opinions of mankind requires that they should declare the causes which impel them to the separation.

"We hold these truths to be self-evident, that all men are created equal, that they are endowed by their Creator with certain unalienable Rights, that among these are Life, Liberty and the pursuit of Happiness.—That to secure these rights, Governments are instituted among Men, deriving their just powers from the consent of the governed,—That whenever any Form of Government becomes destructive of these ends, it is the Right of the People to alter or to abolish it, and to institute new Government."

These sentences, usually called the preamble to the Declaration, laid down the premise upon which the argument for independence was made. The "laws of nature and of nature's God," the "opinions of mankind," and "truths" that were "self-evident" established the basis upon which to judge that the "course of human events" had reached the point where change was both necessary and justified.

As the founding document by the Founding Fathers, the Declaration of Independence has been dissected in every conceivable way and by historians of every stripe. The pertinent phrase is "laws of nature and of nature's God," which sounds, superficially at least, like a reference to Jefferson's favorites, Bacon and Newton.[3] As a scholar of science, one of whose idols was Isaac Newton, Jefferson well understood that one of the goals of natural philosophy was to discover fundamental laws of nature; Newton's three laws of motion were such laws. So too were Boyle's gas law, Kepler's laws of planetary motion, and Archimedes' law of the lever. They were inviolable; they describe the behavior of the material world. So, is that what Jefferson meant by the "laws of nature" in the Declaration of Independence?[4]

At least one historian of science has argued that when Jefferson wrote of the "laws of nature," he was using a direct Newtonian metaphor.[5] Jefferson was saying that the "laws of nature" were as solid and invariant as the laws of motion. Given the ways Newton's laws had been used as a metaphor by others, this interpretation might seem reasonable. It is probably wrong, however. The weight of evidence indicates that for Jefferson the phrase "laws of nature" meant something quite different. Confusion over this matter arises yet again from word usage—in this case the word "nature." "Law of nature" did not, in Jefferson's time, mean (or did not only mean) scientific law.[6]

The concept of the law of nature has a long history whose trajectory started with Aristotle (if not earlier) and leads not to Newton and physics but to constitutional law and theories of society. More than five centuries before Jefferson, Thomas Aquinas had distinguished four kinds of laws: *eternal laws,* the set of laws that control all actions and events in the material universe, the *laws of nature,* which were the moral laws, *divine or revealed laws,* which were the laws dictated by God in the Bible and through his prophets and Jesus, and *human or civil laws.* Eternal laws and the laws of nature were discoverable through the reason with which God had endowed humans. The laws of science, as they came to be discovered, were a part of the eternal laws.

The most influential legal commentator and theorist of Jefferson's time, Sir William Blackstone, defined six kinds of law: *Law as the order of the universe*—arising "when the Supreme Being formed the universe, and created matter out of nothing" and then "impressed certain principles upon that matter, from which it can never depart, and without which it would cease to be. When he put the matter into notion, He established certain laws of motion, to which all moveable bodies must conform." *Law as human action*—the "precepts by which man [uses] both reason and free will . . . in the general regulation of his behaviour." *Law of nature*—the "immutable laws of good and evil, to which the Creator himself in all his dispensations conforms: and which He has enabled human reason to discover, so far as they are necessary for the conduct of human actions." *Revealed law*—"the doctrines . . . found only in the Holy Scriptures." *Law of nations* and *Municipal law*.[7]

"The law of nature," Blackstone also wrote, "being coeval with mankind and dictated by God himself, is of course superior in obligation to any other. It is binding over all the globe, in all countries and at all times: no human laws are of any validity, if contrary to this; and such of them as are valid, derive all their force, mediately or immediately, from this original."[8]

In the hands of a host of European writers, including Locke, Rousseau, Hugo Grotius, and Samuel von Pufendorf—all of them familiar to Jefferson—natural law, or the law of nature, developed into four "branches of theory . . . a theory of Society at large, a theory of the State, a theory of the relations of States . . . and a theory of associations and their relation to the State."[9] The main focus was always the theory of the state.

Blackstone made a firm distinction between any scientific law and the law of nature, or moral law. Civil laws existed not to create rights but to protect existing natural rights. The "primary object of law is to maintain and regulate these absolute rights of individuals." Absolute rights are those "such as would belong to man in a state of nature, and which EVERY MAN is entitled to enjoy, whether in society or out of society." Among such rights were "life and liberty," which "no human legislature may abridge or destroy, unless the OWNER himself shall commit some act that amounts to a forfeiture."[10]

One of the arguments in favor of a Newtonian interpretation of the phrase "laws of nature" is that Jefferson used the plural. The "law of nature," in the singular, would have appeared rhetorically stronger. But Jefferson was combining *two* sets of law, that of nature and that of nature's God, in a single phrase, just as had Blackstone, who had written, "Upon these two foundations, the law of nature and the law of revelation, depend all human laws; that is to say, no human laws should be suffered to contradict these."[11] Since the law of nature was universally considered to be contained within eternal law, in writing of the (law of) nature's God, Jefferson was surely strengthening his case by making an appeal to the revealed law of Blackstone and others.

When it came to rights and laws, Jefferson was heavily influenced by the classics, in addition to Blackstone. He attached a great deal of weight to the Greek philosopher Euripides, who, in the *Phoenissae,* has Jocasta say: "Equality . . . ever linketh friend to friend, city to city, and allies to each other; for Equality is man's natural law; but the less is always in opposition to the greater, ushering in the dayspring of dislike. For it is Equality that hath set up for man measures and divisions of weights and hath distinguished numbers; night's sightless orb, and radiant sun proceed upon their yearly course on equal terms, and neither of them is envious when it has to yield."[12]

The voice of which Jefferson's draft preamble to the Declaration of Independence was the strongest echo, however, was that of John Locke. In his *Second Treatise of Civil Government,* which Jefferson first read as a seventeen-year-old student, Locke wrote that every man in a "state of nature" (that is, "originally and fundamentally") was free. "But though this be a state of liberty, yet it is not a state of license . . . The state of Nature has a law of Nature to govern it, which obliges everyone; and reason, which is that law, teaches all mankind who will but consult it, that, being equal and independent, no one ought to harm another in his life, health, liberty, or possessions."

This, then, was the core of Jefferson's premise for the Declaration of Independence, and in making the case the way he did, he was on very familiar ground. Indeed, some version of the phrase "laws of nature and of nature's God" had appeared in the same connection many times in the previous half-century. The Bostonian lawyer James Otis wrote in 1764, "Should an Act of Parliament be against any of His natural laws [it] would be contrary to eternal truth, equity and justice, and consequently void." Samuel Adams in his *Natural Rights of Colonists as Men* (1772), wrote that "just and true liberty, equally and immortal . . . is a thing that all men are clearly entitled to by the eternal and immutable laws of God and nature."[13] Bolingbroke, the English deist, whom Jefferson admired, had written to the poet Alexander Pope, "I say that the law of nature is the law of God. It is the modest, not the presumptuous, inquirer who makes a real and safe progress in the discovery of divine truths. One follows Nature and Nature's God; that is, he follows God in his works and in his word."[14] Pope himself wrote of the man who "slave to no sect, who takes no private road, but looks through Nature up to Nature's God."[15]

In *A Summary View of the Rights of British America* (1774), Jefferson had written of "a right which nature has given to all men, of departing from the country in which chance, not choice, has placed them," "rights, as derived from the laws of nature, and not as the gift of their chief magistrate," and the "rights of human nature." In "Opinion on the Residence Bill" (1790), Jefferson wrote: "Every man, and every body of men on earth, possesses the right of self-government. They receive it with their being from the hand of nature . . . the law of the *majority* is the natural law of every society of men." And in "Opinion on the Treaties with France" (1793), he wrote: "Those who write treatises of natural law, can only declare what their own moral sense & reason dictate in the several cases they cite. Such of them as happen to have feelings & a reason coincident with those of the wise & honest part of mankind, are respected & quoted as witnesses of what is morally right or wrong in particular cases. Grotius, Pufendorf, Wolf, and Vattel are of this number. Where they agree their authority is strong. But where they differ, & they often differ, we must appeal to our own feelings and reason to decide between them."[16] As late as 1814, when Jefferson was advising on the curriculum for the new University of Virginia, he listed "the Law of Nature and Nations" as a single scholarly subject under the heading of Philosophy and quite distinct from the subjects listed under the "department of Mathematics," within which natural philosophy and the laws of science were now defined as physics.

The quotations in this series summarize Jefferson's view of the law of nature and its relation to natural (equally inviolable) rights. The laws of science as then known (and even now) did not illuminate morals and ethics, although they might be used as metaphors, the way Desaguliers did. Nor do moral laws illuminate scientific problems or produce scientific solutions.

The Declaration of Independence was not a place for experimental philosophical musings or even clever metaphors. Indeed, Jefferson's philosophy required morality to be common to all, through conscience and reason. A vital part of Jefferson's rhetorical plan for the Declaration of Independence was for it to be phrased in terms clear and familiar to all. "Neither aiming at originality of principle or sentiment, nor yet copied from any particular and previous writing, it was intended to be an expression of the American mind, and to give to that expression the proper tone and spirit called for by the occasion. All its authority rests then on the harmonizing sentiments of the day, whether expressed in conversation, in letters, printed essays, or in the elementary books of public right, as Aristotle, Cicero, Locke, Sidney, &c."[17]

It seems clear, therefore, that in this case, we do not see the hand of Jefferson the scientist but the hand of Jefferson the constitutional lawyer. Despite his devotion to the principles and practices of natural philosophy, with all their potential for the improvement of human life, the words "laws of nature and of nature's God" seem to refer neither to science in general nor to Newton in particular. He meant moral law and stated quite clearly, with Euripides and dozens of other philosophers, that moral law entailed the natural rights of equality and freedom ("life, liberty, and the pursuit of happiness"). With Blackstone, he believed that civil law protected those natural rights.

Philosophers may have dreamed that the deep truths of morality would one day be reducible by reason to the laws of the movement of atoms, but Jefferson considered the law of nature to be a separate set of moral laws with which humankind had been endowed by the Creator. As he had written to Peter Carr, "He who made us would have been a pitiful bungler if he had made the rules of our moral conduct a matter of science."[18]

In his original draft of the Declaration of Independence Jefferson had written, "We hold these truths to be sacred and undeniable." The final wording, "We hold these truths to be self-evident," is simpler and more direct. The revision may have been the work of Franklin, but both versions capture the spirit in which Jefferson sought to find the foundations upon which to build a personal philosophy and a nation without giving way to political expediency. His whole life was spent in search of ever more truths that were both sacred and undeniable, whether in philosophy or politics, in law or science.

EPILOGUE

Measuring the Shadow

AS PART OF THE MEASURE OF a complex and elusive man, we may try to sum up the balance sheet of Thomas Jefferson's science. Although he was fascinated by science, we cannot say that he was a scientist in the modern sense; he did not make a living through science, nor did he maintain a scientific laboratory. Nonetheless, he made original contributions to science, and they were highly significant, as was the role that science played in his general thinking.

Daniel Boorstin long ago showed that Jefferson's Enlightenment Age philosophy of nature and man did not survive the onslaught of nineteenth-century materialism. "His morality possessed virtues which a naturalistic morality in America one hundred years later would almost certainly lack."[1] It may well be true that the Darwinian paradigm, industrialization, the search for precious metals and fuels to rip out of the ground, and, today, computers, nuclear energy, and DNA have pushed a Jeffersonian view of nature (and God) to the sidelines. But Jefferson's science was more than philosophy, and while one can argue about his philosophy, he also had significant scientific accomplishments.

If Jefferson's library reflected a conservative and thoroughly Baconian approach to the relationships of the various sciences, in his actions he was decidedly more modern. For him, the traditional line of separation between the observational sciences of natural history and the theoretical sciences of natural philosophy (like mathematics and physics) was blurred more often than it was honored. Finding the identity of the mastodon (the American incognitum), for example, was more than a taxonomical issue; it involved causes, climate, and geography. He used his knowledge of mathematics to design an improved moldboard for a plow. Not only did he keep records of ambient temperatures for fifty years, he explored theories of the possible role of changing land use on the American climate, which he thought was warming.

A surprising amount of Jefferson's science has stood the test of time, and in his writings about race and Indian languages, he proposed theories that were well ahead of their time. On the subject of the color differences between black and white people, he astutely suggested that those differences, whatever their physiological origin, might be reinforced by behavior —each person preferring to find mates, like with like, among those of the same race on the basis of ideals of beauty. It was almost another hundred years before Charles Darwin independently came up with the same idea and termed the process "mate choice and sexual selection."[2]

Jefferson had a similarly pre-Darwinian insight about the origin of diversity among American Indian languages. He wanted to understand why different groups, closely allied in other respects and living next to each other, had different languages and dialects when one might have thought that such differences would be swamped by cultural interchange. The problem is a classic one in evolutionary biology: Why are there so many species instead of just one widespread species? Jefferson suggested that the fragmentation of language groups might be a positive factor in keeping groups distinct and would be accentuated at the borders where groups met. Differences in language would therefore be part of what evolutionists call "isolation mechanisms."

When seen with the unsympathetic eye of hindsight, Jefferson was conservative in some of his science—for instance, by taking a religious view of Creation rather than exploring contemporary views on the origins of mountains and the history of the earth. It is also true, however, that he seemed to leave the door open just a little by allowing that the earth might not have been created in one instant but rather "in time." And that small phrase, ironically, provided an opportunity for critics like Clement Moore to accuse him of not being religious enough.

He was the first to conclude that the mastodon was a different species from the living elephants and that it was adapted for colder weather. He was the first to name and describe a North American fossil mammal (*Megalonyx*). But there was no scientific idea that Jefferson clung to more consistently—perhaps even obstinately—than his denial of extinction, although even here his position was nuanced. He readily accepted that races and populations became extinct, and he was equally sure that God had established mechanisms by which the losses were made up. He owned copies of the evolutionary works of Charles Darwin's grandfather Erasmus Darwin, but he never ventured so far as even to hint at the heresy that new species might arise. In fact, one of his arguments against extinction was the highly logical one that if it occurred at the species level, the world would gradually become unpopulated.

If there was a trend in Jefferson's scientific thinking, it was that he steadily became more disenchanted over time with more hypothetical ideas —whether in earth history or moral philosophy—and more concerned with the practical applications of science to societal problems. And there is no doubting his extraordinary skill as a modifier and creator of mechanical devices, from hygrometers to plows. In parallel, he had a true experimenter's approach to farm and garden and was among the leaders in applying scientific methods to agriculture. Jefferson's view of science and education, especially the application of science to useful ends, effectively led America into the nineteenth century. Ironically, the dream combination of science and education eventually produced an industrialized economy of the sort that he wished to avoid.

Jefferson never stopped being a consummate intellectual; he was perpetually driven by curiosity. In early 1825, at age eighty-one, in one of the last of his wonderful letters to John Adams, Jefferson described a new book that Lafayette had sent him, the results of studies of the brain by the French physiologist Marie-Jean-Pierre Flourens.[3] At this stage of life, others might have been enjoying lighter reading or revisiting favorites of their youth. But Lafayette had known what would catch Jefferson's restless imagination.

Flourens had been trying to find out whether, as various anatomists had been suggesting, the different parts of the brain had different functions. He experimented by cutting into or ablating different parts of the brain in rabbits and pigeons. "He takes out the cerebrum (in the mammalian brain, this is the gray matter) compleatly, leaving the cerebellum and other parts of the system uninjured. The animal loses all it's senses of hearing, seeing, feeling, smelling, tasting, is totally deprived of will, intelligence, memory, perception, etc. yet lives for months in perfect health . . . but without moving . . . He takes the cerebellum (the lower base of the posterior brain) out of others, leaving the cerebrum untouched. The animal retains all it's senses . . . but loses the power of regulated motion, and exhibits all the symptoms of drunkenness . . . a puncture in the medulla oblongata (the brain stem) is instant death."[4]

Jefferson was fascinated to see that Flourens had shown that the cerebrum "is the thinking organ." And this raised the question whether, "if deprived of that organ . . . the soul remains in the body . . . or whether it leaves it as in death, and where it goes."

Perhaps Jefferson was particularly interested in these researches because he knew his own death was imminent. "All this you and I shall know," he told Adams, "when we meet again in another place, and at no distant period." Indeed, they died on the same day—the most fitting day of all—July 4, 1826.

In his career as one of the nation's leading scientific intellectuals, Jefferson helped launch at least four sciences in America: paleontology, climatology, geography, and scientific archaeology. He encouraged the study of many others. *Notes on the State of Virginia* established American natural history as a serious subject. And, remarkably, he accomplished all this—which would have been enough for the lifetime of an ordinary mortal—while designing and building Monticello, creating the University of Virginia and designing its buildings, running a large plantation, and, almost effortlessly, it seems, being not just a political philosopher but a Founding Father and the leader of an entire nation. Of all the things that he accomplished in his life, it is interesting, therefore, that Jefferson insisted that his tomb only record him as:

AUTHOR OF THE DECLARATION OF AMERICAN INDEPENDENCE
OF THE STATUTE OF VIRGINIA FOR RELIGIOUS FREEDOM
AND FATHER OF THE UNIVERSITY OF VIRGINIA

With his brilliant use of language and logic and his passion for both learning and serving, Jefferson followed a career that was a constant search for truths to hold on to and for friends and colleagues with whom to discover and share the wisdom of the world. Sometimes he found himself unable to integrate opposing "truths" into one belief, and that led him into irreconcilable conflicts that confuse and infuriate modern readers. In the story of Jefferson's science, as in the story of so much of his life, there were not always clear winners and losers. Winning arguments was not necessarily where his success lay. Rather, the triumph of Jefferson was in the breadth of his learning, the power of his imagination, his passion for science, and his ability to inspire and lead (and occasionally perplex)—all of which we still celebrate today.

Appendix

Jefferson's Letter on Climate to Jean Baptiste Le Roy

The original letter resides in the Library of Congress. The diagrams are omitted here. Reprinted from The Papers of Thomas Jefferson, *edited successively by Julian P. Boyd, Charles T. Cullen, John Catanzariti, and Barbara B. Oberg (Princeton, NJ: Princeton University Press, 1950–); copyright Princeton University Press.*

Paris, November 13, 1786

Sir, I received the honour of yours of Sep. 18. a day or two after the accident of a dislocated wrist had disabled me from writing. I have waited thus long in constant hope of recovering it's use. But finding that this hope walks before me like my shadow, I can no longer oppose the desire and duty of answering your polite and learned letter. I therefore employ my left hand in the office of scribe, which it performs indeed slowly, awkwardly and badly.

The information given by me to the Marquis de Chastellux, and alluded to in his book and in your letter, was that the sea breezes which prevail in the lower parts of Virginia during the summer months, and in the warm parts of the day, had made a sensible progress into the interior country: that formerly, within the memory of persons living, they extended but little above Williamsburg; that afterwards they became sensible as high as Richmond, and at present they penetrate sometimes as far as the first mountains, which are above an hundred miles farther from the sea coast than Williamsburg is. It is very rare indeed that they reach those mountains and not till the afternoon is considerably advanced. A light North-Westerly breeze is for the most part felt there, while an Easterly, or North Easterly wind is blowing strongly in the lower country. How far Northward and Southward of Virginia this Easterly breeze takes place, I am not informed. I must therefore be understood as speaking of that state only, which extends on the sea coast from 36½ to 38° of latitude.

This is the fact. We know too little of the operations of Nature in the physical world to assign causes with any degree of confidence. Willing always however to guess at what we do not know, I have sometimes indulged myself with conjectures on the causes of the phænomena above stated. I will hazard them on paper for your amusement, premising for their foundation some principles believed to be true.

Air, resting on a heated and reflecting surface, becomes warmer, rarer and lighter: it ascends therefore, and the circumjacent air, which is colder and heavier, flows into it's place, becomes warmed and lightened in it's turn, ascends and is succeeded as that which went before. If the heated surface be circular, the air flows to it from every quarter, like the rays of a circle to it's center. If it be a zone of determinate breadth and indefinite length, the air will flow from each side perpendicularly on it. If the currents of air flowing from opposite sides be of equal force, they will meet in equilibrio at a line drawn longitudinally thro the middle of the zone. If one current be stronger than the other, the stronger one will force back the line of equilibrium towards the further edge of the zone, or even beyond it: the motion it has acquired causing it to overshoot the zone, as the motion acquired by a pendulum in it's descent causes it to vibrate beyond the point of it's lowest descent.

Earth, exposed naked to the sun's rays, absorbs a good portion of them; but being an opaque body, those rays penetrate to a small depth only. It's surface, by this accumulation of absorbed rays, becomes considerably heated. The residue of the rays are reflected into the air resting on that surface. This air then is warmed 1. by the direct rays of the sun. 2. by it's reflected rays. 3. by contact with the heated surface. A Forest receiving the sun's rays, a part of them enter the intervals between the trees, and their reflection upwards is intercepted by the leaves and boughs. The rest fall on the trees, the leaves of which being generally inclined towards the horizon, reflect the rays downwards. The atmosphere here then receives little or no heat by reflection. Again, these leaves having a power of keeping themselves cool by their own transpiration, they impart no heat to the air by contact. Reflection and contact then, two of the three modes before mentioned of communicating heat, are wanting here, and of course the air over a country covered by forest must be colder than that over cultivated grounds. The sea being pellucid, the sun's rays penetrate it to a considerable depth. Being also fluid, and in perpetual agitation, it's parts are constantly mixed together; so that instead of it's heat being all accumulated in it's surface, as in the case of a solid opaque body, it is diffused thro' its whole mass. It's surface therefore is comparatively cool, for these reasons, to which may be added that of evaporation. The small degree of reflection, which might otherways take place is generally prevented by the rippled state of it's surface. The air resting on the sea then, like that resting on a forest, receives little or no heat by reflection or contact; and is therefore colder than that which lies over a cultivated country.

To apply these observations to the phænomena under construction.

The first settlements of Virginia were made along the sea coast, bearing from South towards the North, a little Eastwardly. These settlements formed a zone in which, tho every point was not cleared of it's forest, yet a good proportion was cleared and cultivated. This cultivated earth, as the sun advances above the horizon in the morning, acquires from it an intense heat, which is retained and increased through the warm parts of the day. The air resting on it becomes warm in proportion and rises. On one side is a country still covered with forest: on the other is the ocean. The colder air from both of these then rushes towards the heated zone to supply the place left vacant there by the ascent of it's warm air. The breeze from the West is light and feeble; because it traverses a country covered with mountains and forests, which retard it's current. That from the East is strong; as passing over the ocean wherein there is no obstacle to it's motion. It is probable therefore that this Easterly breeze forces itself far into, or perhaps beyond the zone which produces it. This zone is, by the increase of population, continually widening into the interior country. The line of equilibrium between the Easterly and Westerly breezes is therefore progressive.

Did no foreign causes intervene, the sea breezes would be a little Southwardly of the East, that direction being perpendicular to our coast. But within the tropics there are winds which blow continually and strongly from the East. This current affects the courses of the air even without the tropics. The same cause too which produces a strong motion of the air from East to West between the tropics, to wit, the Sun, exercises it's influence without these limits, but more feebly in proportion as the surface of the globe is there more obliquely presented to it's rays. This effect, tho' not great, is not to be neglected when the sun is in, or near, our summer solstice, which is the season of these Easterly breezes.

The Northern air too, flowing towards the equatorial parts to supply the vacuum made there by the ascent of their heated air, has only the small rotatory motion of the polar latitudes from which it comes. Nor does it suddenly acquire the swifter rotation of the parts into which it enters. This gives it the effect of a motion opposed to that of the earth, that is to say of an Easterly one. And all these causes together are known to produce currents of air in the Atlantic, varying from East to North East as far as the 40th. degree of Latitude. It is this current which presses our sea breeze out of it's natural South Easterly direction to an Easterly and sometimes almost a North Easterly one.

We are led naturally to ask where the progress of our sea breezes will ultimately be stopped? No confidence can be placed in any answer to this question. If they should ever pass the Mountainous country which separates the waters of the Ocean from those of the Missisipi, there may be circumstances which might aid their further progress as far as the Missisipi. That Mountainous country commences about 200 miles from the sea coast, and consists of successive ranges, passing from North East to South West, and rising the one above the other to the Alleghaney ridge, which is the highest of all. From that, lower and lower ridges succeed one another again till, having covered in the whole a breadth of 200 miles from South East to North West, they subside into plain, fertile country, extending

400 miles to the Missisipi, and probably much further on the other side towards the heads of it's Western waters. When this country shall become cultivated, it will, for the reasons before explained, draw to it winds from the East and West. In this case, should the sea breezes pass the intermediate mountains, they will rather be aided than opposed in their further progress to the Missisipi. There are circumstances however which render it possible that they may not be able to pass those intermediate mountains. 1. These mountains constitute the highest lands within the United States. The air on them must consequently be very cold and heavy, and have a tendency to flow both to the East and West. 2. Ranging across the current of the sea breezes, they are in themselves so many successive barriers opposed to their progress. 3. The country they occupy is covered with trees, which assist to weaken and spend the force of the breezes. 4. It will remain so covered; a very small proportion of it being capable of culture. 5. The temperature of it's air then will never be softened by culture.

At present I suppose the currents of air between the Atlantic and the Western heads of the Missisipi may be represented as in the following diagram of a horizontal section of that country. But that when the plane country on both sides of the Missisipi shall be cleared of it's trees and cultivated, the currents of air will be in the following directions.

Whether, in the plane country between the Missisipi and Alleganey mountains Easterly or Westerly winds prevail at present, I am not informed. I conjecture however that they must be Westerly, as represented in the first diagram: and I think, with you Sir, that if those mountains were to subside into plane country as their opposition to the Westerly winds would then be removed, they would repress more powerfully those from the East, and of course would remove the line of equilibrium nearer to the sea-coast for the present.

Having had occasion to mention the course of the Tropical winds from East to West, I will add some observations connected with them. They are known to occasion a strong current in the ocean in the same direction. This current breaks on that wedge of land of which Saint Roque is the point; the Southern column of it *probably* turning off and washing the coast of Brazil. I say *probably* because I have never heard the fact and conjecture it from reason only. The Northern column, having it's Western motion diverted towards the North and reinforced by the currents of the great rivers Orinoko, Amazons and Tocantin, has probably been the agent which formed the gulph of Mexico, cutting the American continent nearly in two in that part. It re-crosses into the ocean at the Northern end of the gulph, and passes, by the name of the Gulph stream, all along the coast of the United States to it's Northern extremity. There it turns off Eastwardly, having formed, by it's eddy at this turn, the banks of New found land. Thro' the whole of it's course, from the gulph to the banks, it retains a very sensible warmth. The Spaniards are at this time desirous of trading to their Philippine islands by the way of the Cape of good hope: but opposed in it by the Dutch, under authority of the treaty of Munster, they are examining the practicability of a common passage thro' the Streights of Magellan, or round Cape Horn. Were they to make an opening thro the isthmus of Panama, a work much less difficult than some even of the inferior canals of France, however small this opening should be in the beginning, the tropical current, entering it with all it's force, would soon widen it sufficiently for it's own passage, and thus complete in a short time that work which otherwise will still employ it for ages. Less country too would be destroyed by it in this way. These consequences would follow. 1. Vessels from Europe, or the Western coast of Africa, by entering the tropics, would have a steady wind and tide to carry them thro' the Atlantic, thro America and the Pacific ocean to every part of the Asiatic coast, and of the Eastern coast of Africa: thus performing with speed and safety the tour of the whole globe, to within about 24°. of longitude, or 1/15 part of it's circumference, the African continent, under the line, occupying about that space. 2. The gulph of Mexico, now the most dangerous navigation in the world, on account of it's currents and moveable sands, would become stagnant and safe. 3. The gulph stream on the coast of the United States would cease, and with that those derangements of course and reckoning which now impede and endanger the intercourse with those states. 4. The fogs on the banks of Newfoundland*, supposed to be the vapours of the gulph stream rendered turbid by cold air, would disappear. 5. Those banks, ceasing to receive supplies of sand, weeds and warm water by the gulph stream, it might become problematical what effect changes of pasture and temperature would have on the fisheries. However it is time to relieve you from this lengthy lecture. I wish it's subject may have been sufficiently interesting to make amends for it's details. These are submitted with entire deference to your better judgment. I will only add to them by assuring you of the sentiments of perfect esteem and respect with which I have the honor to be Sir your most obedient and most humble servant,

Th. Jefferson

*This ingenious and probable conjecture I find in a letter from Dr. Franklin to yourself published in the late volume of the American Philosophical Transactions.

Notes

Abbreviations

In the notes, where certain key works are referenced multiple times, they are cited using the following abbreviations.

Adams-Jefferson	*The Adams-Jefferson Letters: The Complete Correspondence between Thomas Jefferson and Abigail and John Adams.* Edited by Lester J. Cappon. Chapel Hill: University of North Carolina Press, 1959.
Autobiography	Thomas Jefferson. *Autobiography.* In *The Life and Selected Works of Thomas Jefferson,* edited by Adrienne Koch and William Peden. New York: Modern Library, 1944.
Boyd	*The Papers of Thomas Jefferson.* Edited successively by Julian P. Boyd, Charles T. Cullen, John Catanzariti, and Barbara B. Oberg. Princeton, NJ: Princeton University Press, 1950–.
Buffon	Georges-Louis Leclerc, Comte de Buffon. *Histoire Naturelle, Générale et Particulière, avec la Description du Cabinet du Roi.* Paris: Imprimerie Royale, 1744–1788. Page references are given for the original quarto edition. The translations are mine unless specifically cited from William Smellie, *A Natural History, General and Particular . . . ,* translated from the French (London: Evans, 1817).
Ford	*The Works of Thomas Jefferson.* Edited by Paul Leicester Ford. New York: G. P. Putnam's Sons, 1904–1905.
Garden Book	*Thomas Jefferson's Garden Book.* Edited by Edwin Morris Betts. Philadelphia: American Philosophical Society, 1985.
L&B	*The Writings of Thomas Jefferson.* Edited by Andrew A. Lipscomb and Albert Ellery Bergh. Washington, DC: Thomas Jefferson Memorial Foundation, 1903–1904.

LOC	The Thomas Jefferson Papers, 1606–1827. Library of Congress, Washington, DC. Available online at http://memory.loc.gov/ammem/collections/jefferson_papers/.
Memorandum Books	*Jefferson's Memorandum Books: Accounts, with Legal Records and Miscellany, 1767–1826.* Edited by James A. Bear Jr. and Lucia C. Stanton. Princeton, NJ: Princeton University Press, 1997.
Notes	Thomas Jefferson. *Notes on the State of Virginia.* Edited by Frank Shuffelton. New York: Penguin, 1999. All citations are to this Penguin Classics edition.
RS	*The Papers of Thomas Jefferson, Retirement Series.* Edited by J. Jefferson Looney. Princeton, NJ: Princeton University Press, 2004–. These volumes cover the years from 1809 on.

Introduction

1. The present work is a direct extension of my previous books *The Legacy of the Mastodon* (New Haven: Yale University Press, 2008) and *A Passion for Nature* (Chapel Hill: University of North Carolina Press; Charlottesville, VA: Thomas Jefferson Memorial Foundation, 2008) introducing Jefferson's science to the general reader.
2. TJ to John Melish, December 10, 1814, *L&B* 14, pp. 219–220.
3. Charles A. Miller, *Jefferson and Nature* (Baltimore, MD: Johns Hopkins University Press, 1993).
4. TJ to Samuel Kercheval, July 14, 1816, *L&B* 15, pp. 323–344, inscribed on the Jefferson Memorial, Washington, DC.

CHAPTER ONE. Lost: One Large Moose

1. TJ to Buffon, October 1, 1787, *Boyd* 12, pp. 194–195.
2. John Sullivan to TJ, April 16, 1787, *Boyd* 11, p. 295.
3. The debate over state and federal jurisdiction and responsibility, which began with the Constitution of 1787 and was intensely debated after Jefferson and James Madison wrote the Kentucky and Virginia Resolutions of 1798, continues today over matters such as social security and national medical insurance.

21. TJ to Daniel Salmon, February 15, 1808, *L&B* 11, pp. 440–441.
22. TJ to Peter Carr, August 10, 1787, *Boyd* 12, pp. 14–15.
23. Daniel J. Boorstin, *The Lost World of Thomas Jefferson* (Boston: Beacon Press, 1948).
24. Jefferson's Memorandum Book entry for May 3, 1773, shows that he paid David Jameson, the treasurer, his fee of eighteen shillings. The following year, on June 15, he paid twenty shillings. *Memorandum Books* 1, p. 446.
25. Jefferson's name first appeared in the minutes of the American Philosophical Society in late 1779 recording receipt of "letters from the Revd. Mr. Maddison, President of William & Mary's College in Virginia containing a Series of meteorological observations by his Excellency Governor Jefferson and himself seperately for an year and a half; likewise a set of Experiments on what are called the 'sweet springs.' Mr. Rittenhouse is desired to thank both those gentlemen in the name of the Society for these favors, and to request a continuance of their valuable correspondence." Manuscript Minutes, December 17, 1779, American Philosophical Society, Philadelphia.

CHAPTER FOUR. A Measured and Orderly World

1. *Garden Book,* pp. 16–17.
2. Keith Thomson, *A Passion for Nature* (Chapel Hill: University of North Carolina Press and Thomas Jefferson Memorial Foundation, 2008), pp. 10–11.
3. See, for example, Jan Golinski, *British Weather and the Climate of Enlightenment* (Chicago: University of Chicago Press, 2007).
4. Deists rejected the alternative approach, which was to understand God through his revelations.
5. TJ to B. Vaughan, July 23, 1788, *Boyd* 15, pp. 394–398.
6. TJ to James Madison, April 4, 1792, *Boyd* 27, pp. 818–822.
7. TJ, "Second State of the Report on Weights and Measures," *Boyd* 16, pp. 628–648.
8. Rhys Isaac, *Landon Carter's Uneasy Kingdom* (Oxford: Oxford University Press, 2004), p. 77.
9. The Reverend Madison sent some detailed observations for 1779 to David Rittenhouse at the American Philosophical Society. They were published in 1789 in the second volume of the society's *Transactions.*
10. TJ to Mary Jefferson, June 13, 1789, *Boyd* 16, pp. 491–492.
11. Mary Jefferson to TJ, August 20, 1790, *Boyd* 17, pp. 332–333.
12. TJ to Maria Jefferson, March 9, 1791, *Boyd* 19, p. 427.
13. The six children of Thomas Jefferson and Martha Wales Skelton Jefferson were

Martha (1772–1836), Jane (1774–1775), an unnamed son (1777), Mary ("Polly," "Maria," 1778–1809), Lucy Elizabeth (1780–1782), and Elizabeth (1782–1785).

14. Barbé-Marbois sent a set of these queries to senior people in each of the thirteen states.
15. *Autobiography,* p. 94.
16. Perhaps Jefferson kept a copy of his report to Barbé-Marbois, but it and the original are missing.
17. "Gen. Sullivan's Description of New Hampshire to the Marquis de Marbois," in *Letters and Papers of Major-General John Sullivan, Continental Army,* ed. Otis G. Hammond (Concord, NH, 1939), vol. 3, pp. 229–239.
18. The first edition of *Notes* was printed in France by Philippe-Denis Pierre in May 1785. When a pirated French edition appeared, Jefferson had a new edition published by John Stockdale in London in 1787.
19. *Notes,* pp. 68–69.

CHAPTER FIVE. Science and the Mastodon

1. Ezra Stiles, quoted in *Boyd* 7, p. 302. Emphasis in the original.
2. *Philosophical Transactions of the Royal Society of London* 29 (1714), pp. 62–71.
3. We now know that the remains are around ten thousand years old.
4. Keith Thomson, *The Legacy of the Mastodon* (New Haven: Yale University Press, 2008); Stanley Hedeen, *Big Bone Lick* (Lexington: University Press of Kentucky, 2008); Paul Semonin, *American Monster* (New York: New York University Press, 2000).
5. George Cuvier, "Sur le Grande Mastodonte, Animal Très-Voisin de l'Eléphant, mais à Mâchelières Hérissées de Gros Tubercles, Dont on Trouve les Os en Divers Endroits des Deux Continens, et surtout près des Ords de l'Ohio, dans l'Amérique Septentrionale Improprement Nommé Mammouth par les Anglais et par les Habitans des Etats-Unis," *Paris, Annales du Muséum d'Histoire Naturelle* 8 (1806), pp. 270–312.
6. William Hunter, "Observations on the Bones Commonly Supposed to Be Elephant's Bones, Which Have Been Found near the River Ohio, in America," *Philosophical Transactions of the Royal Society of London* 58 (1769), pp. 34–45.
7. Benjamin Franklin to Abbé Chappe d'Auteroche, January 31, 1768, *The Papers of Benjamin Franklin,* ed. Leonard W. Larabee (New Haven: Yale University Press, 1959–), vol. 15, pp. 33–34.
8. *Buffon,* Supplementary Volume 5 (1778), pp. 545–552.
9. Buffon's discussions of the remains from Big Bone Lick were not purely

empirical; they were strongly conditioned by his philosophy of zoology. For him the elephant and hippopotamus were single species within which variations were seen. Because he saw limits to that variability, he could not accept the most obvious conclusion about the remains from Big Bone Lick: that the huge tusks, the giant limb bones, and the massive teeth all belonged to a single animal. He could not countenance an "elephant" that did not have "elephant-like" teeth or, indeed, a hippopotamus with more than four cusps in the molar teeth. And he had difficulty accepting that there could have been forms of either that were massively larger than the living forms. His philosophy led him to an interesting conclusion: the animal to which the massive teeth belonged must now be extinct because something so large would have been seen by now if it were still alive.

10. TJ's analysis of the mastodon is in *Notes,* pp. 43–49.
11. Circular Letter of 1797, published in 1799: *Transactions of the American Philosophical Society* 4 (1799), pp. xxxvii–xxxix.
12. Robert Livingston to TJ, January 7, 1801, *Boyd* 32, pp. 406–408.
13. Thomson, *Legacy of the Mastodon,* pp. 46–56.
14. Georges Cuvier eventually became the authority on the mastodon, giving it the name "mastodonte" and concluding that it was a different species from the Siberian mammoth and very definitely extinct.

CHAPTER SIX. The Natural History of Virginia and America

1. Charles Thomson to TJ, March 9, 1782, *Boyd* 6, pp. 163–164.
2. The only tapir species in North America is Baird's tapir, *Tapirus bairdii,* which ranges from Mexico to Panama and may weigh more than 800 pounds; Jefferson listed a weight of 534 pounds. The wild boar of Europe may reach 650 pounds; Jefferson listed it at 280.
3. See, for example, TJ to Thomas Walker, September 25, 1783, *Boyd* 6, pp. 339–340.
4. TJ to William Whipple, January 12, 1784, *Boyd* 27, pp. 735–736, and April 27, 1784, *Boyd* 27, p. 739; John Sullivan to TJ, March 12, 1784, *Boyd* 7, pp. 21–24; TJ to John Sullivan, April 27, 1784, *Boyd* 7, pp. 317–321.
5. James Madison to TJ, May 12, 1786, *Boyd* 9, pp. 517–522.
6. Robert Trow-Smith, *A History of British Livestock Husbandry, 1700–1900* (London: Routledge and Paul, 1959).
7. Peter Kalm, *Travels into North America: Containing Its Natural History . . . with the Civil, Ecclesiastical and Commercial State of the Country . . .* (1770), trans. John Reinhold Foster (London: Imprint Society, 1978).

8. William Byrd, *Natural History of Virginia, 1737,* ed. Richard Croom Beatty and William Mulloy (Richmond, VA: Dietz, 1940), p. 18. The case has been made that the real author was a land developer named Samuel Jenner; see Percy G. Adams, "The Real Author of William Byrd's *Natural History of Virginia,*" *American Literature* 28 (1956), pp. 211–220.
9. *The Private Correspondence of Daniel Webster,* ed. Fletcher Webster (Boston: Little, Brown, 1875), vol. 1, p. 371. In Jefferson's library there was a copy of Buffon's "Époques de la Nature" (Supplementary Volume 5 [1778] of his *Histoire Naturelle*) inscribed to Jefferson from Buffon.
10. Keith Thomson, *A Passion for Nature* (Chapel Hill: University of North Carolina Press and Thomas Jefferson Memorial Foundation, 2008), pp. 66–70; Thomson, "Jefferson, Buffon, and the European View of America," in *The Libraries, Leadership, and Legacy of John Adams and Thomas Jefferson,* ed. Robert C. Baron and Conrad Edick Wright (Golden, CO: Fulcrum; Boston: Massachusetts Historical Society, 2010), pp. 225–244.
11. TJ to Buffon, October 1, 1787, *Boyd* 12, pp. 194–195.
12. Mark Catesby, *The Natural History of Carolina, Florida and the Bahama Islands: Containing the Figures of Birds, Beasts, Fishes, Serpents, Insects, and Plants: Particularly the Forest-Trees, Shrubs, and Other Plants, Not Hitherto Described, or Very Incorrectly Figured by Authors. Together with Their Descriptions in English and French. To Which, Are Added Observations on the Air, Soil, and Waters: with Remarks upon Agriculture, Grain, Pulse, Roots, &c. To the Whole, Is Prefixed a New and Correct Map of the Countries Treated Of* (London, 1731–1743).
13. Jeremy Mynott, in *Birdscapes: Birds in Our Imagination and Experience* (Princeton, NJ: Princeton University Press, 2009), gives an edited version of Jefferson's bird lists.
14. Leaves lanceolate, pointed, serrated, etc.
15. The first American to publish a work using Linnaean names was William Young Jr.: *Catalogue d'Arbres, Arbustes et Plantes Herbacées d'Amérique* (Paris, 1783; facsimile reprint by Samuel N. Rhoads, Philadelphia, 1916).
16. TJ to John Manners, February 22, 1814, *L&B* 14, p. 98.
17. E. Millicent Sowerby, *Catalogue of the Library of Thomas Jefferson* (Charlottesville: University of Virginia Press, 1983), vol. 1, p. 460.
18. In the 1785 letter to Henley, he remarked how difficult it would be for anyone to locate them in his library at Monticello.
19. Henley's was obviously another important library. Jefferson bought seven works by Linnaeus: two regional botanies, *Flora Japponica* and *Fauna Suecica* (Finland), together with *Critica Botanica, Genera Plantarum, Species Plan-*

tarum, "*Emantissa altera*" (incorrect titles for *Mantissa Plantarum Altera Generum*), and finally the great classic *Systema Naturae,* along with fifty-one other titles. Among the last were John Clayton's *Flora Virginica (Gronovius)* and Thomas Whately's *Observations on Modern Gardening.* And, in yet another demonstration of his catholic tastes, he also bought books on medals, poetry and music, three volumes of Dante, a history of dueling, a chemical dictionary, and a book on Anglo-Saxon coins.

20. TJ to Henley, June 9, 1778, *Boyd* 2, pp. 198–199. Henley had the same sort of bad luck with his library as, eventually, did Jefferson. Henley's books were damaged in a flood at Madison's house in 1778. In a letter of March 3, 1785, Jefferson congratulated himself on having taken the books he wanted before "that general destruction which involved the residue of your books when Mr. Madison's house was burnt" (in 1781). TJ to Henley, *Boyd* 8, pp. 11–14.
21. *Buffon* 9 (1778), pp. 106–110, 1761 (trans. Smellie 12, pp. 204–259).
22. Ibid., p. 258.

CHAPTER SEVEN. Mountains and Shells

1. Samuel Dexter, "A Letter on the Retreat of House-Swallows in Winter," *Memoirs of the American Academy of Arts and Sciences* 1 (1785), pp. 494–496. The hypothesis that swallows and other species hibernated rather than migrating owed its origin to Aristotle, if not to someone earlier. It was a long time a-dying. People were sure that they had observed swallows enter and leave the water. Even the British naturalist the Reverend Gilbert White, of Selborne fame, was reluctant to give up the idea, although his brother wrote to him from Gibraltar reporting that he had seen flocks of springtime migrants overhead, heading north.
2. Genesis, chapter 9, King James version.
3. John C. Greene, *The Death of Adam* (Ames: Iowa State University Press, 1959); Keith Thomson, *Before Darwin* (New Haven: Yale University Press, 2005).
4. Thomas Burnet, *Telluris Theoria Sacra (Sacred History of the Earth)* (London, 1681).
5. Leonardo da Vinci had long since examined these questions in his then unpublished notebooks. "Why are the bones of great fishes, and oysters and corals and various other shells and sea-snail,—shells all intermingled, which have become part of stone—[why are they] found on the high tops of mountains?" He concluded that if seashells were found on mountains, then the seas had once covered those mountains. The shells had not been carried into the

mountains after the animals' death, "because the years of their growth are numbered upon the outer coverings of their shells; and both small and large ones may be seen; and these would not have grown without feeding, or fed without growing, . . . and here they would not have been able to move." As for the idea that the shells were merely simulacra produced by the rocks: "If you say that these shells have been and still are being created in such places by the nature of the locality or by the potency of the heavens in these spots, such an opinion cannot exist in brains possessed of any extensive powers of reasoning." It must rather be the case that "the Peaks of the Apennines once stood up in a sea, in the form of islands surrounded by salt water, and above the plains of Italy where flocks of birds are flying today, fishes were once moving in the large shoals." J. P. Richter, *The Literary Works of Leonardo da Vinci* (Cambridge: Cambridge Universty Press, 1939), vol. 2, p. 175.

6. For example, William Whiston, *A New Theory of the Earth from Its Original, to the Consummation of All Things . . .* (London, 1697).
7. Nicolas Steno [Niels Stenson], *De Solido Intra Solidum Naturaliter Contento Dissertation Prodromus* (Florence, 1669).
8. Robert Hooke, "Lectures and Discourses of Earthquakes and Subterraneous Eruptions," in *The Posthumous Works of Robert Hooke, MD, FRS,* ed. Richard Walker (London: Royal Society of London, 1709).
9. *Notes,* pp. 32–33. The shells found in Virginia were probably Silurian age brachiopod shells.
10. Rev. James Madison to TJ, December 28, 1786, *Boyd* 10, p. 643.
11. TJ to David Rittenhouse, January 25, 1786, *Boyd* 9, pp. 215–217.
12. A fine selection of these can be found in Robert Plot, *The Natural History of Oxford-shire* (Oxford, 1677).
13. Charles Thomson to TJ, July 8, 1786, *Boyd* 10, pp. 102–105.
14. TJ to Charles Thomson, December 17, 1786, *Boyd* 10, pp. 608–610.
15. TJ to Charles Thomson, April 20, 1787, *Boyd* 12, pp. 59–166.
16. The modern answer is that both forces had been at work in the earth, but the earth—and granites—had indeed first been molten.
17. TJ to C. F. C. de Volney, February 8, 1805, *L&B* 11, pp. 62–69.
18. TJ to Dr. John P. Emmett, May 2, 1826, *L&B* 16, pp. 168–172.
19. James Hutton, *Abstract of a Dissertation . . . Concerning the System of the Earth, Its Duration, and Stability* (Royal Society of Edinburgh, 1785).
20. Benjamin Franklin, "Conjectures Concerning the Formation of the Earth," *Transactions of the American Philosophical Society* 3 (1793), pp. 1–5.

CHAPTER NINE. Europe and the Peoples of America

1. *Buffon* 9 (1778), pp. 104–105 (trans. KST).
2. TJ to John Adams, June 11, 1812, *Adams-Jefferson,* pp. 305–308.
3. Jefferson's observations on American Indians are all in his chapter entitled "Aborigines" in *Notes,* pp. 98–113.
4. *The Commonplace Book of Benjamin Rush,* in *The Autobiography of Benjamin Rush: His "Travels through Life" Together with His Commonplace Book for 1789–1813;* "Travels through Life" edited with an introduction and notes by George W. Corner (Princeton, NJ: Published for the American Philosophical Society by Princeton University Press, 1948).
5. Hugh Jones, The *Present State of Virginia* (London, 1724), p. 11.
6. Mark Catesby, *The Natural History of Carolina, Florida and the Bahama Islands . . .* (London, 1731–1743).
7. In later editions of *Notes,* Jefferson added as an appendix numerous letters supporting the charge that Logan's family had been murdered by Cresap and his party. "Logan" was an adopted English name; his real name may have been Tah-ga-jute, and he was not technically a chief.
8. TJ to John Adams, June 11, 1812, *Adams-Jefferson,* pp. 305–308.
9. Anthony F. C. Wallace, *Jefferson and the Indians* (Cambridge, MA: Harvard University Press, 1999).
10. See ibid., chapter 8, for example.
11. Keith Thomson, "Jefferson, Buffon, and the European View of America," in *The Libraries, Leadership, and Legacy of John Adams and Thomas Jefferson,* ed. Robert C. Baron and Conrad Edick Wright (Golden CO: Fulcrum; Boston: Massachusetts Historical Society, 2010), pp. 225–244.
12. *Buffon,* Supplementary Volume 5 (1778), p. 237.
13. C. Vann Woodward, *The Old World's New World* (Oxford: Oxford University Press, 1991). See also Henry Steele Commager and Elmo Giordanetti, *Was America a Mistake?* (New York: Harper Torchbooks, 1967).
14. Hannah Adams, *A Summary History of New England* (Dedham, MA, 1799), p. 230.
15. TJ to Chevalier de Chastellux, June 7, 1785, *Boyd* 8, pp. 184–186.
16. De las Casas, in his *Brevísima relación de la destrucción de las Indias,* stated: "They are by nature the most humble, patient, and peaceable, holding no grudges, free from embroilments, neither excitable nor quarrelsome. These people are the most devoid of rancors, hatreds, or desire for vengeance of any people in the world. And because they are so weak and complaisant, they are less able to endure heavy labor and soon die of no matter what malady. The

sons of nobles among us, brought up in the enjoyments of life's refinements, are no more delicate than are these Indians, even those among them who are of the lowest rank of laborers."

17. William Robertson, *The History of America* (London: Cadell and Davies, 1812), vol. 1, pp. 318–319.
18. Antonio D'Ulloa and Jorge Juan, *Travels to South America* (London, 1772), pp. 420, 269, 402.
19. Ibid., p. 21.
20. Ibid., p. 468.
21. Catesby, *Natural History,* 8.
22. D'Ulloa and Juan, *Travels to South America,* pp. 478–479.
23. TJ to Benjamin Vaughan, July 23, 1788, *Boyd* 15, pp. 394–398.
24. Robertson, *History of America,* vol. 2, p. 199.
25. Ibid., p. 63.
26. Antonio D'Ulloa, *Noticias Americanas, Entretenimiento XVIII* (Madrid, 1771), pp. 267–268.
27. Antonello Gerbi's *The Dispute of the New World* (English version: Pittsburgh: University of Pittsburgh Press, 1973) is a classic of intellectual history writing, reaching far beyond the ambit of the present work.
28. Peter Kalm, *Travels into North America: Containing Its Natural History . . . with the Civil, Ecclesiastical and Commercial State of the Country . . .* (1770), trans. John Reinhold Foster (1770; London: Imprint Society, 1978), p. 59.
29. Ibid.
30. Carmichael to TJ, October 15, 1787, *Boyd* 12, p. 241.
31. *Notes,* pp. 69–70.
32. John Adams to TJ, February 1814, *Adams-Jefferson,* pp. 426–430.
33. Buffon, "Sur les Amèricains," in *Buffon,* Supplementary Volume 4 (1777), pp. 525–539.
34. Edward Seeber, "Chief Logan's Speech in France," *Modern Language Notes* 16 (1946), pp. 214–416; Dwight Boehm and Edward Schwartz, "Jefferson and the Theory of Degeneracy," *American Quarterly* 9 (1957), pp. 448–453.
35. Benjamin Franklin, "Observations Concerning the Increase of Mankind," in *The Papers of Benjamin Franklin,* ed. Leonard W. Larabee (New Haven: Yale University Press, 1968), vol. 4, pp. 225–234.
36. *Buffon,* Supplementary Volume 4, p. 526.
37. See James Madison to TJ, December 10, 1783, *Boyd* 6, pp. 377–378.
38. Jefferson wrote: "In a later edition of the Abbé Raynal's work, he has withdrawn his censure from that part of the new world inhabited by the Federo-Americans; but has left it still on the other parts. North America has always

been more accessible to strangers than South. If he was mistaken then as to the former, he may be so as to the latter. The glimmerings which reach us from South America enable us only to see that its inhabitants are held under the accumulated pressure of slavery, superstition, and ignorance. Whenever they shall be able to rise under this weight, and to shew themselves to the rest of the world, they will probably shew they are like the rest of the world."

CHAPTER TEN. Natural History, Slavery, and Race

1. Annette Gordon-Reed, "Thomas Jefferson and St. George Tucker: The Makings of Revolutionary Slaveholders," in *Jefferson, Lincoln and Wilson: The American Dilemma of Race and Democracy,* ed. John Milton Cooper Jr. and Thomas J. Knock (Charlottesville: University of Virginia Press, 2010), pp. 15–33; Winthrop Jordan, *White over Black* (Williamsburg, VA: Omohundro Institute, 1968).
2. *Notes,* pp. 168–169.
3. TJ to James Madison, June 17, 1785, *Boyd* 8, pp. 227–234.
4. Charles Thomson to TJ, November 2, 1785, *Boyd* 9, pp. 9–10.
5. Jefferson had written positively about slavery in ancient Rome; he retained this passage, over Thomson's objections.
6. A half-chance for Jefferson to emancipate his own slaves was presented in the form of a legacy from the war hero Thaddeus Kosciusko that was to pay for it. However, Jefferson dithered, and the money was never released. See Gary Nash and Graham Hodges, *Friends of Liberty: A Tale of Three Patriots, Two Revolutions, and the Betrayal That Divided a Nation—Thomas Jefferson, Thaddeus Kosciuszko, and Agrippa Hull* (New York: Basic Books, 2008).
7. Lord Kames, *Six Sketches of the History of Man* (London, 1776), book 1, p. 46.
8. *Buffon* 3 (1749), pp. 468–469, and Supplementary Volume 4 (1777), pp. 502–505.
9. *Notes,* p. 146.
10. Phillis Wheatley, *Poems on Various Subjects, Religious and Moral* (Boston: Bell, 1773).
11. Jefferson had an amusing exchange of correspondence with John Adams on the lost tribe: TJ to John Adams, June 11, 1812, *Adams-Jefferson,* pp. 305–308.
12. *Buffon* 3, pp. 498–499.
13. Samuel Stanhope Smith, *An Essay on the Causes of the Variety of Complexion and Figure in the Human Species,* 2nd ed. (New York: Williams and Whiting, 1810).
14. *Notes,* p. 146.

15. The scarf skin is the epidermis.
16. *Notes,* pp. 145–146.
17. *Buffon* 14 (1766), pp. 44–45. In Buffon's time, there was some confusion about how many kinds of great apes there were and where they lived. Buffon recognized two kinds, "the pongo" and "the jocko" (or "organ-outan"), but he used the term "organg-outang" for the apes that lived in Africa, when it is now clear that travelers had been referring to the chimpanzee or baboon. (The gorilla was not discovered until 1846.) Buffon evidently enjoyed stories of simian debauchery. "Dampier, Froger, and other travelers, assure us, that the orang-outans carry off girls of eight or ten years of age to the tops of trees, and that it is extremely difficult to rescue them. To these testimonies we may add that of M. de la Brosse, who assures us, in his voyage to Angola, in the year 1733, that the orang-outans, which he calls quimpezés, endeavour to surprise the Negresses, whom they detain for the purpose of enjoying them, and entertain them plentifully." Ibid.
18. As described by Thomas Schaeper in *Edward Bancroft: Scientist, Author, Spy* (New Haven: Yale University Press, 2011), Bancroft turns out to be one of the most interesting, if reprehensible, characters of Jefferson's acquaintanceship.
19. See A. Glenn Crothers, "Quaker Merchants and Slavery in Early National Alexandria, Virginia," *Journal of the Early Republic* 25 (2005), pp. 47–77.
20. Here is the act in full. "Virginia Legislature, May 1782—ACT XXI. An act to authorize the manumission of slaves. Be it therefore enacted, That it shall hereafter be lawful for any person, by his or her last will and testament, or by any other instrument in writing, under his or her hand and seal, attested and proved in the county court by two witnesses, or acknowledged by the party in the court of the county where he or she resides, to emancipate and set free, his or her slaves, or any of them, who shall thereupon be entirely and fully discharged from the performance of any contract entered into during servitude, and enjoy as full freedom as if they had been particularly named and freed by this act.

    "That all slaves so set free, not being in the judgment of the court, of sound mind and body, or being above the age of forty-five years, or being males under the age of twenty-one, or females under the age of eighteen years, shall respectively be supported and maintained by the person so liberating them, or by his or her estate; and upon neglect or refusal so to do, the court of the county where such neglect or refusal may be, is hereby empowered and required, upon application to them made, to order the sheriff to distrain and sell so much of the person's estate as shall be sufficient for that purpose."
21. Edward Bancroft to TJ, September 16, 1788, *Boyd* 13, pp. 606–608. This is

not the place to discuss the complicated reasons why this might be so, if indeed it was true. In a later letter to Jefferson, his friend the financier William Short outlined a similar plan that he thought of trying himself. William Short to TJ, October 7, 1793, *Boyd* 27, p. 203.

22. TJ to Bancroft, January 26, 1789, *Boyd* 14, pp. 492–494.
23. Peter S. Onuf, "Domesticating the Captive Nation: Thomas Jefferson and the Problem of Slavery," in *Seeing Jefferson Anew, in His Time and Ours,* ed. John B. Boles and Randal I. Hall (Charlottesville: University of Virginia Press, 2010), pp. 13–29.
24. *Notes,* pp. 144–145.
25. *Autobiography,* p. 77.
26. Douglas R. Egerton, *Gabriel's Rebellion: The Virginia Slave Conspiracies of 1800 and 1802* (Chapel Hill: University of North Carolina Press, 1993).
27. Andrew Levy, *The First Emancipator* (New York: Random House, 2005).
28. Peter S. Onuf, "Thomas Jefferson and the Problem of Slavery," in *Jefferson, Lincoln and Wilson: The American Dilemma of Race and Democracy,* ed. John Milton Cooper Jr. and Thomas J. Knock (Charlottesville: University of Virginia Press, 2010), pp. 34–60.
29. TJ to Henri Gregoire, February 25, 1809, *Ford* 9, pp. 246–247.
30. TJ to Banneker, August 30, 1791, *Boyd* 22, p. 97.
31. The story originated with Jefferson's grandson Thomas Jefferson Randolph and appears in his manuscript *Memoirs* (1874), in the Edgehill-Randolph papers at the University of Virginia, accession number 1397: "On riding with Mr. Jefferson when president we met a Negro who bowed to us. Mr. J returned the salute, I did not; he turned to me after we had passed and asked if I permitted a Negro to be more of a gentleman than myself." A similar story is associated with both James Madison and George Washington. Information from Anna Berkes, Monticello Library.
32. TJ to Joel Barlow, October 8, 1809, *Ford* 9, pp. 261–265. Andrew Ellicott, not as well known for his part in planning Washington, DC, was a distinguished Philadelphia mathematician, astronomer, and surveyor.
33. TJ to Edward Coles, August 25, 1814, *Ford* 9, pp. 477–479.

CHAPTER ELEVEN. The Color of Their Skin

1. Annette Gordon-Reed, *The Hemingses of Monticello* (New York: Norton, 2008).
2. TJ to Edward Coles, August 25, 1814, *Ford* 9, pp. 477–479. This quotation continues the passage at the end of Chapter 9.
3. *Notes,* chapter 14 ("Laws").

4. *Buffon,* Supplementary Volume 4 (1777), pp. 556–570 (Smellie 10, p. 421).
5. Henry Skipwith to TJ, January 20, 1784, *Boyd* 6, pp. 472–474.
6. Charles Carter to TJ, February 9, 1784, *Boyd* 6, pp. 534–535.
7. *Notes,* pp. 77–78.
8. The cause of albinism is a genetic mutation. The cause of vitiligo is unknown, but it may be connected with an auto-immune disease like alopecia, hyperthyroidism, or pernicious anemia.
9. Buffon, "Sur les Blasards et Nègres Blancs," in *Buffon,* Supplementary Volume 4 (1777), pp. 555–578; see also Buffon, "Variétés dans l'Espèce Humaine," in *Buffon* 3, pp. 371–530.
10. William Byrd, "An Account of a Negro-Boy That Is Dappel'd in Several Places of His Body with White Spots," *Philosophical Transactions of the Royal Society of London* 19 (1657), pp. 781–782.
11. Alexander Russel, "An Account of a Remarkable Alteration of Colour in a Negro Woman: In a Letter to the Reverend Mr. Alexander Williamson of Maryland, from James Bate in That Province," *Philosophical Transactions of the Royal Society of London* 51 (1759), pp. 175–178.
12. John Morgan, "Some Account of a Motley Coloured, or Pye Negro Girl and Mulatto, Exhibited before the Society on the Month of May, 1784," *Transactions of the American Philosophical Society* 2 (1786), pp. 392–395. Charles Willson Peale gave a similar report concerning a man from Maryland: "Account of a Black Man Turned White," in *The Selected Papers of Charles Willson Peale,* ed. Lillian B. Miller (New Haven: Yale University Press, 1983), vol. 1, pp. 620–621.
13. Broadside, in *The Commonplace Book of Benjamin Rush,* in *The Autobiography of Benjamin Rush: His "Travels through Life" Together with His Commonplace Book for 1789–1813;* "Travels through Life" edited with an introduction and notes by George W. Corner (Princeton, NJ: Published for the American Philosophical Society by Princeton University Press, 1948).
14. Benjamin Smith Barton's "Account of Henry Moss, a White Negro," *Philadelphia Medical and Physical Journal,* was originally read to the American Philosophical Society on September 16, 1796. Benjamin Rush saw Moss on July 27 and recorded similar notes in his Commonplace Book.
15. Rev. James Madison to TJ, December 28, 1786, *Boyd* 10, pp. 642–644. Emphasis in the original—perhaps Madison was countering Jefferson's statement about the (inferior) lack of redness in albinos.
16. *Transactions of the American Philosophical Society* 4 (1799), pp. 289–297. Jefferson's paper on the fossil he called the great-claw was published in the same volume.

17. *Commonplace Book of Benjamin Rush.* This was the African Methodist Episcopal Church of Richard Allen, the first black church in Philadelphia.

CHAPTER TWELVE. The Paris Years

1. William Henry de Saussure, *An Address to the Citizens of South-Carolina on the Approaching Election of President and Vice-President of the United States* (Charleston, 1800), p. 10.
2. TJ to Joseph Willard, March 24, 1789, *Boyd* 14, pp. 697–699.
3. Gaye Wilson, "'Behold Me at Length on the Vaunted Scene of Europe': Jefferson and the Creation of an American Image Abroad," in *Old World, New World: America and Europe in the Age of Jefferson,* ed. Leonard J. Sadosky, Peter Nicolaisen, Peter S. Onuf, and Andrew J. O'Shaughnessy (Charlottesville: University of Virginia Press, 2010).
4. Jon Kukla, *Mr. Jefferson's Women* (New York: Knopf, 2008); compare Virginia Scharff, *The Women Jefferson Loved* (New York: HarperCollins, 2010).
5. Jan Ellen Lewis, "Jefferson and Women," in *Seeing Jefferson Anew: In His Time and Ours,* ed. John B. Boles and Randal L. Hall (Charlottesville: University of Virginia Press, 2010), pp. 152–171.
6. TJ to James Monroe, May 21, 1784, *Boyd* 7, p. 279.
7. Sidney I. Pomerantz, "George Washington and the Inception of Aeronautics in the Young Republic," *Proceedings of the American Philosophical Society* 90 (1954), pp. 131–138.
8. TJ to Philip Turpin, April 28, 1784, *Boyd* 7, pp. 134–137. See Barthélemy Faujas de Saint-Fond, *Description des Experiences de la Machine Aerostatique* (Paris, 1783–1784).
9. TJ, "Note on Balloons," circa January 9, 1793, *Boyd* 25, p. 42.
10. TJ to James Madison, June 17, 1785, *Boyd* 8, pp. 227–234.
11. TJ to Philip Turpin, April 28, 1784, ibid.
12. TJ, "Notes from Condorcet on Slavery," no date, *Boyd* 14, pp. 494–498.
13. TJ to George Fleming, December 31, 1815, *L&B* 14, pp. 365–369.
14. There is a drawing (N566.M16) in the Coolidge collection of Jefferson papers at the Massachusetts Historical Society that is labeled (in a later hand) "steam engine," but consultation with experts like Jeff Lock and Henry Petroski convinces me that it was not such. What the drawing *does* represent remains a mystery.
15. Even so, Boulton's London mill was a financial failure, perhaps because it consumed one hundred bushels of coal for each day's operation.
16. TJ to Charles Thomson, December 17, 1786, *Boyd* 10, pp. 608–610.

17. TJ to Charles Thomson, December 17, 1786, *Boyd* 10, pp. 608–610.
18. TJ to Thomas Paine, April 3, 1789, *Boyd* 14, pp. 372–377.
19. Robert Livingston to TJ, January 26, 1799, *Boyd* 30, pp. 653–656.
20. TJ to Rev. James Madison, October 2, 1785, *Boyd* 8, pp. 574–577.
21. TJ to Hugh Williamson, February 6, 1785, *Boyd* 7, pp. 641–643.
22. TJ to George Washington, July 17, 1785, *Boyd* 8, p. 301; Washington to TJ, September 26, 1785, in *The Papers of George Washington,* ed. William Wright Abbot, Dorothy Twohig, Philander D. Chase, Beverly H. Runge, and Frederick Hall Schmidt (Charlottesville: University of Vriginia Press, 1983–1995), vol. 8, pp. 555–558.
23. TJ to Rev. William Smith, 1791, *Boyd* 19, pp. 112–113.
24. TJ to Charles Thomson, October 8, 1785, *Boyd* 8, p. 598.
25. Charles Thomson to TJ, March 6, 1785, *Boyd* 8, pp. 15–17.
26. TJ to David Rittenhouse, November 11, 1784, *Boyd* 7, pp. 516–518.
27. TJ to Rev. James Madison, October 2, 1785, *Boyd* 8, pp. 374–376; Madison to TJ, March 27, 1786, *Boyd* 9, pp. 355–357.
28. TJ to Rev. James Madison, July 19, 1788, *Boyd* 13, pp. 379–383.
29. *Early Proceedings of the American Philosophical Society for the Promotion of Useful Knowledge; Compiled by One of the Secretaries; from the Manuscript Minutes of Its Meetings from 1744 to 1838* (1884), pp. 126–127.
30. TJ to Hugh Williamson, February 6, 1785, *Boyd* 7, p. 641.
31. Charles Thomson to TJ, March 6, 1785, *Boyd* 8, p. 17.
32. TJ to David Rittenhouse, January 25, 1786, *Boyd* 9, p. 215.
33. TJ to Joseph Willard, March 24, 1789, *Boyd* 14, pp. 687–699.
34. TJ to Rev. James Madison, July 19, 1788, *Boyd* 13, pp. 379–382.
35. TJ to Thomas Cooper, July 10, 1812, *L&B* 13, pp. 176–178.
36. TJ to Thomas Jefferson Randolph, January 3, 1808, quoted in E. Millicent Sowerby, *Catalogue of the Library of Thomas Jefferson* (Charlottesville: University of Virginia Press, 1983), vol. 1, p. 374.
37. John Whately, *Observations on Modern Gardening* (London, 1770).
38. TJ, "Notes of a Tour into the Southern Parts of France, &c., 3 March–10 June 1787," *Boyd* 11, pp. 415–464.
39. TJ to Madame de Tessé, March 20, 1787, *Boyd* 11, pp. 226–228.
40. TJ to John Jay, May 4, 1787, *Boyd* 8, pp. 338–390.
41. TJ, "Notes of a Tour into France," pp. 460–462.
42. TJ, "Notes of a *Tour* through *Holland* and the Rhine Valley," *Boyd* 13, pp. 8–33.
43. TJ to John Jay, August 30, 1785, *Boyd* 8, pp. 452–455.
44. TJ to Henry Knox, September 12, 1789, *Boyd* 15, pp. 422–423.

CHAPTER THIRTEEN. The Practical Scientist

1. Silvio A. Bedini, *Thomas Jefferson: Statesman of Science* (New York: Macmillan, 1990). Bedini was the unquestioned expert on Jefferson's mechamical devices.
2. Nathaniel P. Poor, *Catalogue: President Jefferson's Library* (Washington, DC: Gales and Seaton, 1829).
3. TJ to John P. Emmet, in *The Writings of Thomas Jefferson,* ed. H. A. Washington (New York: Derby and Jackson, 1859), vol. 7, pp. 441–444.
4. Susan R. Stein, *The World of Thomas Jefferson at Monticello* (Charlottesville, VA: Thomas Jefferson Memorial Foundation, 1993).
5. TJ to William Short, February 11, 1789, *Boyd* 14, pp. 538–543; TJ, undated note "Maccaroni," *Boyd* 14, p. 544.
6. TJ to James Madison, February 8, 1786, *Boyd* 9, pp. 264–267.
7. TJ to Benjamin Vaughan, July 23, 1788, *Boyd* 13, pp. 394–398.
8. Benjamin Franklin, "Letter to Mr. Nairne, of London, from Dr. Franklin, Proposing a Slowly Sensible Hygrometer for Certain Purposes," *Transactions of the American Philosophical Society* 2 (1759), pp. 51–56.
9. TJ to Benjamin Vaughan, December 29, 1786, *Boyd* 10, pp. 646–647.
10. TJ to Charles Willson Peale, March 21, 1815.
11. TJ, "Notes of a *Tour* through *Holland* and the Rhine Valley," *Boyd* 13, pp. 9–33.
12. TJ to John Taylor, December 29, 1794, *Boyd* 28, pp. 230–234.
13. TJ to Benjamin Latrobe, *Latrobe Papers* 3 (1817), pp. 808–809.
14. Bedini, *Thomas Jefferson;* Martin Clagett, *Scientific Jefferson Revealed* (Charlottesville: University of Virginia Press, 2009).

CHAPTER FOURTEEN. Climate and Geography

1. Edward Augustus Holyoke, "A Table of Results, from a Course of Observations Made on the Heat of the Atmosphere, by Fahrenheit's Thermometer," *Memoirs of the American Academy of Arts and Sciences* 3 (1793), pp. 89–92.
2. Madison to TJ, January 22, 1784, *Boyd* 6, pp. 507–508.
3. *Memorandum Books* 1, p. 420.
4. *Memorandum Books* 1, p. 49.
5. William A. Koelsch, "Thomas Jefferson, American Geographers, and the Uses of Geography," *Geographical Review* 98 (2008), pp. 260–279.
6. In 1779, Madison sent a year's run of observations to David Rittenhouse for publication by the American Philosophical Society: *Transactions of the American Philosophical Society* 2 (1786), pp. 141–158.

7. TJ to Rev. James Madison, December 24, 1783; letter no longer extant, see *Boyd* 6, p. 420.
8. Rev. James Madison to TJ, January 22, 1784, *Boyd* 6, pp. 507–508.
9. TJ to Thomas Mann Randolph, April 18, 1790, *Boyd* 16, pp. 350–352.
10. TJ to James Madison, February 20, 1784, *Boyd* 6, pp. 544–546; the town of Orange is some twenty miles northeast of Charlottesville.
11. TJ to James Madison, March 16, 1784, *Boyd* 7, pp. 30–31.
12. Susan Solomon, John S. Daniel, and Daniel Druckenbrod, "Revolutionary Minds," *American Scientist* 95 (2007), pp. 430–437.
13. Dunbar sent Jefferson sets of meteorological data that were published by the American Philosophical Society in 1809, together with a cover letter mentioning western fossil remains: "Extract from a Letter, from William Dunbar, of the Natchez, to Thomas Jefferson, President of the Society," *Transactions of the American Philosophical Society* 6 (1809): 40–42.
14. TJ to Giovanni Fabbroni, June 8, 1778, *Boyd* 2, pp. 195–198.
15. In this chapter, all the quotations from *Notes on the State of Virginia* are from chapter 7, "Climate."
16. TJ to Benjamin Vaughan, July 23, 1788, *Boyd* 13, pp. 394–398; see also TJ to Vaughan, January 26, 1787, *Boyd* 11, pp. 69–72.
17. Vaughan to TJ, August 2, 1788, *Boyd* 13, pp. 459–461.
18. TJ, "Notes of a Tour into the Southern Parts of France, &c., 3 March–10 June 1787," *Boyd* 11, p. 442.
19. TJ to C. F. C. de Volney, February 8, 1805, *LOC*.
20. TJ to Dr. N. Chapman, December 11, 1809, *L&B* 12, pp. 338–339. "The year of the deep snow" corresponds with one of the temperature minima of the Little Ice Age that followed the Medieval Warm Period of global climate.
21. Jefferson's "Summary of His Meteorological Journal for the Years 1810 through 1816 at Monticello," in *Garden Book*, pp. 623–624.
22. Kenneth Thompson, "The Question of Climatic Stability in America before 1900," *Climate Changes* 3 (1981), pp. 227–241.
23. Benjamin Franklin to Ezra Stiles, May 29, 1763, *The Papers of Benjamin Franklin*, ed. Leonard W. Larabee (New Haven: Yale University Press, 1959–), vol. 10, pp. 264–265.
24. Hugh Williamson, "An Attempt to Account for the Change of Climate, Which Has Been Observed in the Middle Colonies in North-America," *Transactions of the American Philosophical Society* 1 (1770), pp. 272–280.
25. Thomas Wright, "On the Mode Most Easily and Effectually Practicable of Drying Up the Marshes of the Maritime Parts of North America," *Transactions of the American Philosophical Society* 4 (1799), pp. 243–246.

26. Samuel Williams, *Natural and Civil History of Vermont* (Walpole, NH, 1794), p. 70.
27. G. K. van Hogendorp to TJ, September 8, 1785, *Boyd* 3, pp. 501–503.
28. Noah Webster, "A Dissertation on the Supposed Changes of the Temperature of Winter," *Memoirs of the Connecticut Academy of Arts and Sciences* 1 (1810), pp. 1–68.
29. Alexander von Humboldt, *Ansichten der Nature,* rev. ed. (1808; Stuttgart and Tübingen: Cotta, 1849).
30. F. J. de B. Chevalier de Chastellux, *Travels in North America in the Years 1780, 1781, and 1782,* ed. Howard C. Rice (Chapel Hill: University of North Carolina Press, 1963), vol. 2, pp. 395–396.
31. B. Henry Latrobe and William Tatham, "Memoir on the Sand-Hills of Cape Henry in Virginia," *Transactions of the American Philosophical Society* 4 (1799), pp. 438–444.
32. TJ to Jean Baptiste Le Roy, November 13, 1786, *Boyd* 10, pp. 524–530.
33. David M. Ludlum, "Thomas Jefferson and the American Climate," *Bulletin of the American Meteorological Society* 47 (1966), pp. 974–975.
34. TJ to George Hopkins, September 5, 1822, *L&B* 15, pp. 394–395.
35. TJ to Lewis E. Beck, July 16, 1824, *L&B* 13, pp. 71–73.

CHAPTER FIFTEEN. Redeeming the Wilderness

1. *Notes,* p. 101.
2. See Gilbert Chinard, "The American Philosophical Society and the Early History of Forestry in America," *Proceedings of the American Philosophical Society* 89 (1945), pp. 444–488; "Eighteenth Century Theories on America as a Human Habitat," ibid., 91 (1947), pp. 27–57.
3. John Adams to Abigail Adams, August 14, 1776, *Adams Family Correspondence,* vol. 2, ed. L. H. Butterfield (Cambridge, MA: Belknap Press of Harvard University Press, 1963), p. 96.
4. The classic treatment of this subject is Roderick Frazier Nash, *Wilderness and the American Mind,* 4th ed. (New Haven: Yale University Press, 1982).
5. J. Hector St. John de Crèvecoeur, *Letters from an American Farmer,* ed. Albert E. Stone (1781; New York: Penguin, 1986).
6. Hugh Jones, *The Present State of Virginia* (London: Clarke,1724), p. 35.
7. William Cronon, *Changes in the Land* (New York: Hill and Wang, 1983), pp. 142–143.
8. Andrea Wulf, *The Brother Gardeners* (London: Knopf, 2008).
9. TJ to G. K. van Hogendorp, October 13, 1785, *Boyd* 8, pp. 633–634.
10. David Howell to William R. Stapes, quoted in *Boyd* 4, pp. 585–586.

11. *A Guide to the Wilderness; or, The History of the First Settlements in Western Counties of New York, with Useful Instructions to Future Settlers, in a Series of Letters Addressed by Judge Cooper of Coopers-Town to William Sampson, Barrister of New York* (Dublin: Gilbert and Hodges, 1810), pp. 10–11.
12. "Answer of the Lord Culpepper to the Several Articles of His Lordship's Instruction, 1683," quoted in Pierre Marambaud, *William Byrd of Westover, 1674–1744* (Charlottesville: University of Virginia Press, 1971), p. 172.
13. TJ to Chevalier de Chastellux, September 2, 1785, *Boyd* 8, pp. 468–469.
14. TJ to van Hogendorp, October 13, 1785, *Boyd* 8, pp. 633–634.
15. Keith Thomson, *A Passion for Nature* (Chapel Hill: University of North Carolina Press and Thomas Jefferson Memorial Foundation, 2008), p. 36.
16. TJ to Thomas Mann Randolph, January 1, 1792, *Boyd* 23, pp. 7–8.
17. TJ to Madame de Tessé, January 30, 1803, quoted in *Garden Book,* p. 285.
18. TJ to William Prince, July 6, 1791, *Boyd* 20, pp. 603–604.
19. TJ to Bernard McMahon, January 3, 1810, *RS* 2, p. 140.
20. TJ to John Dortic, October 1, 1811, *RS* 5, p. 176.
21. *Memorandum Books* 1, p. 27.
22. Isaac Jefferson, "Memoirs of a Monticello Slave," in *Jefferson at Monticello,* ed. James A. Bear Jr. (Charlottesville: University of Virginia Press, 1967).
23. TJ to George Washington, June 28, 1793, *Boyd* 26, pp. 396–398.
24. George Washington to Arthur Young, August 6, 1786, *The Papers of George Washington,* ed. Theodore J. Crackel (Charlottesville: University of Virginia Press, 2008), vol. 4, pp. 196–210. Between 1790 and 1793, Washington and Arthur Young engaged in a fascinating correspondence comparing the economics of farming in Britain and Virginia, with Jefferson also adding commentary; see TJ, June 18, 1793, "Notes on Mr. Young's Letter," *Boyd* 24, pp. 95–98.
25. Arthur Young, *Rural Oeconomy; or, Essays on the Practical Parts of Husbandry,* American edition (Philadelphia, 1776).
26. Noah Webster to George Washington, July 28, 1790, *Papers of George Washington,* vol. 5, pp. 133–142.
27. TJ to James Madison, June 29, 1793, *Boyd* 26, pp. 401–404.
28. TJ to William Strickland, March 23, 1798, *Boyd* 30, pp. 209–213.
29. August John Foster, 1807, in *Visitors to Monticello,* ed. Merrill D. Peterson (Charlottesville: University of Virginia Press, 1989), p. 39.
30. Landon Carter, "Observations Concerning the Fly-Weevil, That Destroys the Wheat, with Some Useful Discoveries and Conclusions, Concerning the Propagation and Progress of That Pernicious Insect, and the Methods to Be Used for Preventing the Destruction of the Grain by It," *Transactions of the American Philosophical Society* 1 (1769), pp. 205–217.

31. American Philosophical Society, *Circular on the Hessian Fly April 17, 1791, Boyd* 14, pp. 430–432.
32. TJ to Judge William Johnson, May 10, 1817, quoted in *Garden Book,* p. 572.

CHAPTER SIXTEEN. The Unknown West

1. Robert J. Miller, *Native America, Discovered and Conquered: Thomas Jefferson, Lewis and Clark, and Manifest Destiny* (Westport CT: Praeger, 2006), p. 72.
2. TJ to Albert Gallatin, January 7, 1808, *L&B* 11, pp. 415–416.
3. *Garden Book,* p. 95.
4. All the instructions to the expedition in this chapter are from *Letters of the Lewis and Clark Expedition with Related Documents,* 2nd ed., ed. Donald Jackson (Urbana: University of Illinois Press, 1978).
5. John A. Moody, Robert H. Meade, and David R. Jones, *Lewis and Clark's Observations and Measurements of Geomorphology and Hydrology, and Changes with Time,* U.S. Geological Survey Circular 1246 (2003).
6. Lewis to TJ, May 18, 204, *Letters of the Lewis and Clark Expedition,* p. 192.
7. Earl Spamer, R. M. McCourt, R. Middleton, E. Gilmore, and S. B. Duran, "A National Treasure: Accounting for the Natural History Specimens from the Lewis and Clark Expedition (Western North America, 1803–1806) in the Academy of Natural Sciences of Philadelphia," *Proceedings of the Academy of Natural Sciences of Philadelphia* 150 (2000), pp. 47–58. The specimen was from an extinct genre of fishes.
8. *Notes,* pp. 107–108.
9. Andrea Wulf, *The Brother Gardeners* (London: Knopf, 2008).
10. They were entrusted by the American Philosophical Society to the Academy of Natural Sciences of Philadelphia and are stored in its Herbarium.
11. Paul Russell Cutright, *Lewis and Clark, Pioneering Naturalists* (Lincoln: University of Nebraska Press, 1969).
12. Castle McLaughlin. *Arts of Diplomacy: Lewis and Clark's Indian Collection* (Seattle: University of Washington Press, 2003); Anthony F. C. Wallace, *Jefferson and the Indians* (Cambridge, MA: Harvard University Press, 1999), p. 106.
13. Keith Thomson, *A Passion for Nature* (Chapel Hill: University of North Carolina Press and Thomas Jefferson Memorial Foundation, 2008), pp. 16–120.
14. TJ to Bernard McMahon, January 6, 1807, in *Letters of the Lewis and Clark Expedition,* pp. 356–357.
15. Peter Hatch, "'Public Treasures'": Thomas Jefferson and the Garden Plants of Lewis and Clark," *Twinleaf Journal Online* (2003), available at http://www.monticello.org/.

16. Bernard McMahon, *Account of Louisiana, Being an Abstract of Documents Delivered in, or Transmitted to, Mr. Jefferson, President of the United States of America and by Him Laid before Congress, and Published by Their Order. Printed at Washington, and Reprinted at Philadelphia, and All the Other States of the Union* (London: Reprinted for John Hatchard, 1804).
17. Nicholas Biddle, *History of the Expedition under the Command of Captains Lewis and Clark, to the Sources of the Missouri* (Philadelphia: Allen, 1814); Frederick Pursh, *Flora Americae Septentrionalis* (London: Cochrane, 1814).

CHAPTER SEVENTEEN. "His Theories I Cannnot Admire"

1. Francis Hopkinson to TJ, March 20, 1785, *Boyd* 8, p. 52.
2. From this point, it is all downhill. It continues with this stanza:

   Besides an hundred wonders more,
   Of which we never heard before.
   But now, dear doctor, not to flatter,
   There is a most important matter,
   A matter which you never touch on,
   A subject if we right conjecture,
   Well deserves a long, long lecture,
   Which all the ladies would approve—
   The Natural History of Love.

   This piece of doggerel is perhaps notable only because Byron felt moved to pen an equally awful poem to answer it. (Some even think Byron might have been the author of both, but the honor for the original may go to Elizabeth Hamilton of Edinburgh.) Interestingly, despite, or perhaps because of, the popularity and accessibility of Dr. Moyse's book, Jefferson did not own a copy.
3. Quoted in Harry H. Clark, "The Influence of Science on American Ideas," *Transactions of the Wisconsin Academy of Sciences, Arts, and Letters* 35 (1943), pp. 305–349.
4. John Adams to Abigail Adams, May 12, 1770, *Adams Family Correspondence,* vol. 3, ed. L. H. Butterfield (Cambridge, MA: Belknap Press of Harvard University Press, 1973), p. 342.
5. *Diary and Autobiography of John Adams,* ed. L. H. Butterfield (Cambridge, MA: Belknap Press of Harvard University Press, 1961), vol. 1, p. 260.
6. Ibid., p. 62.

7. Adams to Benjamin Waterhouse, April 23, 1785, in *Statesman and Friend: Correspondence of John Adams with Benjamin Waterhouse,* ed. Worthington Chauncey Ford (Boston: Little, Brown, 1927), p. 7.
8. *Diary and Autobiography of John Adams,* vol. 2, p. 22: April 24, 1756.
9. *Earliest Diary and Autobiography of John Adams,* ed. L. H. Butterfield (Cambridge, MA: Belknap Press of Harvard University Press, 1966), p. 76.
10. Ibid.
11. *Diary and Autobiography of John Adams,* vol. 1, p. 71.
12. Adams, quoted in Edward Handler, "Nature Itself Is All Arcanum: The Scientific Outlook of John Adams," *Transactions of the American Philosophical Society* 120 (1976), pp. 216–229.
13. Ibid.
14. John Adams to Francis Adrian van der Kemp, November 5, 1804, *Adams Papers, 1639–1889,* microfilm edition (Boston: Massachusetts Historical Society, 1954–1959), reel 118.
15. Adams to van der Kemp, July 13, 1801, ibid.
16. Adams to van der Kemp, January 8, 1806, ibid.
17. *Diary and Autobiography of John Adams,* vol. 2, pp. 336–337.
18. Adams to van der Kemp, February 19, 1814, *Adams Papers,* reel 95.
19. TJ to Daniel Salmon, February 15, 1808, *L&B* 11, pp. 440–442.
20. Adams to TJ, June 28, 1812, *Adams-Jefferson,* p. 182.
21. TJ to Peter Carr, August 10, 1787, *Boyd* 12, p. 14.
22. *Earliest Diary and Autobiography of John Adams,* March 2, 1756.
23. Marginalium in Adams's copy of Winthrop's Lecture on Earthquakes (1758), *Earliest Diary and Autobiography of John Adams,* pp. 61–62.
24. *Diary and Autobiography of John Adams,* vol. 2, p. 139.
25. JA to Waterhouse, February 24, 1791, idem.
26. JA to Benjamin Waterhouse, September 8, 1784, in *Statesman and Friend.*
27. Adams to van der Kemp, November 10, 1801, and van der Kemp's replies on December 30, 1801, and January 3, 1802, *Adams Papers,* reel 118.
28. *Diary and Autobiography of John Adams,* vol. 2, pp. 245–246.

CHAPTER EIGHTEEN. Philosophers Unwelcome

1. Anonymous in *Port Folio,* quoted in *The Atlantic Monthly* (1859), p. 10.
2. Clement C. Moore, *Observations upon Certain Passages in Mr. Jefferson's Notes on Virginia* (New York, 1804), pp. 30–31.
3. Timothy Dwight, *The Duty of Americans, at the Present Crisis, Illustrated in a Discourse Preached on the Fourth of July, 1798* (New Haven, CT, 1798), p. 10.

4. Ibid., p. 20.
5. James E. Lewis, "What Is to Become of Our Government?" in *The Revolution of 1800,* ed. James Horn, Jan Ellen Lewis, and Peter S. Onuf (Charlottesville: University of Virginia Press, 2002).
6. Timothy Dwight, *A Discourse on Some Events of the Last Century* (New Haven, CT, 1801), pp. 20–21. America was then in the midst of a religious revival of which Dwight was a leader.
7. See, for example, Stephen Waldman, *Founding Faith: Providence, Politics and the Birth of Religious Freedom in America* (New York: Random House, 2008).
8. TJ to John Adams, April 11, 1823, *Adams-Jefferson,* pp. 591–594.
9. Edwin S. Gaustad, *Sworn on the Altar of God: A Religious Biography of Thomas Jefferson* (Grand Rapids, MI: Eerdmans, 1996). Most of the Founding Fathers were more conventional Christians; see Herbert M. Morais, *Deism in Eighteenth Century America* (New York: Russell and Russell, 1960).
10. Thomas Paine, 1797, "Of the Religion of Deism Compared with the Christian Religion. 1804," in *The Life and Writings of Thomas Paine,* ed. Daniel E. Wheeler (Parke: New York, 1808).
11. *Jefferson's Literary Commonplace Book,* ed. Douglas L. Wilson (Princeton, NJ: Princeton University Press, 1989), pp. 42–43.
12. *Jefferson's Extracts from the Gospels,* ed. Dickinson W. Adams (Princeton, NJ: Princeton University Press, 1983). Eugene Sheridan's introductory essay for this volume was republished, with additions, as *Jefferson and Religion* (Charlottesville, VA: Thomas Jefferson Foundation, 1998).
13. The formal title was "A Visit from Saint Nicholas."
14. Moore, *Observations.*
15. Luther Martin, "Letter to Mr. James Fennel," *Philadelphia Gazette,* March 30, 1797; and printed contemporaneously in other periodicals, such as *Porcupines' Gazette.*
16. John Adams to Benjamin Rush, July 17, 1803, *Adams Papers, 1639–1889,* microfilm edition (Boston: Massachusetts Historical Society, 1954–1959), reel 118.
17. John Adams to William Tudor, January 16, 1800, in Page Smith, *John Adams* (New York: Doubleday, 1962), vol. 2, p. 1034.
18. John Adams to Benjamin Waterhouse, February 26, 1817, in *Statesman and Friends,* pp. 122–125.
19. TJ to John Adams, June 15, 1813, *Adams-Jefferson,* pp. 331–333.
20. John Adams to TJ, June 25, 1813, *Adams-Jefferson,* pp. 333–335. The separation and balance of powers in government was a favorite theme of Adams, and he bolstered it with his knowledge of static and dynamic equilibria, learned at

Harvard. But the concept did not *originate* with physics or theory. It was a longstanding factor in English government, traceable back at least as far as the Magna Charta, which established a tripartite balance of powers among crown, nobility, and commons. For a while, England even experimented with a two-part government, when the monarchy was abolished in the English Civil War. Physicists would note with pleasure that the two-part government turned out to be an unstable arrangement. The apparent inequality of the contending political powers even offered an occasion to invoke Archimedes' law of the lever (i.e., magnitudes are in equilibrium at distances reciprocally proportional to their weights). In government, a few powerful men near the seat of power can be balanced by a weaker cohort farther away.

21. William Henry de Saussure, *An Address to the Citizens of South-Carolina on the Approaching Election of President and Vice-President of the United States* (Charleston, 1800), pp. 18–19.
22. Anonymous, *The Jonny-Cake Papers* (1879).
23. Dwight, *Duty of Americans,* p. 20.
24. TJ, Second Annual Message to Congress, December 15, 1803, *Ford* 8, p. 181.
25. TJ to Edward Jenner, May 14, 1806, *L&B* 19, p. 152.
26. TJ to Benjamin Waterhouse, June 26, 1801, *Boyd* 34, p. 462; Address of Black Hoof, February 5, 1802, *Boyd* 36, pp. 517–522 (see the extensive editorial notes to this address).
27. In 1791, Jacob Isaack of Rhode Island claimed a reward from Congress for an improved method of desalinating seawater—long a subject of great importance for sailors. Issack claimed that adding a secret mixture to the seawater improved the efficiency of distilling freshwater from it. Jefferson commissioned Rittenhouse, Wistar, and James Hutchinson (professor of chemistry at the University of Pennsylvania) to conduct experiments, and they proved the claim to be false. Affidavit of the Secretary of State on the results of the experiments, March 26, 1791, *Boyd* 19, pp. 617–619. Typically, Jefferson then researched the whole history of the subject and sent a second report: "Report on the Desalination of Sea Water," November 21, 1791, *Boyd* 22, pp. 318–322.
28. The United States Naval Academy was not established until 1845.
29. *Memoirs of John Quincy Adams,* ed. Charles Francis Adams (Philadelphia: Lippincott, 1874–1877), p. 317.
30. Ibid., p. 331.
31. Ibid., pp. 472–473. For an appreciation of Madison's ideas on land use, see Andrea Wulf, *Founding Gardeners* (New York: Knopf, 2010).
32. *Memoirs of John Quincy Adams,* p. 317.

33. Jefferson, "A Bill for the More General Diffusion of Knowledge," *Boyd* 2, pp. 526–535.
34. TJ, "A Bill for Amending the Constitution of the College of William and Mary, and Substituting More Certain Revenues for Its Support," *Boyd* 2, pp. 535–545.
35. Thomas Perkins Abernethy, *Historical Sketch of the University of Virginia* (Richmond, VA: Dietz Press, 1948); Cameron Addis, *Jefferson's Vision for Education, 1760–1845* (New York: Lang, 2003); Jennings I. Waggoner, *Jefferson and Education* (Chapel Hill: University of North Carolina Press, 2004).
36. TJ to Peter Carr, September 7, 1814, in *The Life and Selected Writings of Thomas Jefferson,* ed. Adriènne Kock and William Peden (New York: Modern Library, 1944), pp. 642–649. He recommended a separate (mathematics) professor for "Chemistry, Zoology, Botany, Mineralogy." This reflects both the growing importance of chemistry and his continuing appreciation of the details (mineralogy) rather than the theories of geology.

CHAPTER NINETEEN. Transcendental Truths

1. Most authors of Newtonian metaphors referred to offsetting "forces." Voltaire, for example, in his extremely influential work introducing Newton to a French audience (and then English speakers in the English translation), wrote that each planet as it orbits the sun "is under two forces; it tends in a line . . . and gravitates at the same time toward the sun." In fact, technically speaking, orbiting planets do not represent a balance of two offsetting *forces* (force being the acceleration of a mass). In such cases, one factor, the gravitational force, acts in concert with the planet's *momentum* (due to its mass and linear velocity) which was sometimes called its projectile force, to produce a regularity of motion.
2. J. T. Desaguliers, *The Newtonian System of the World, the Best Model of Government: An Allegorical Poem* (London, 1728). The extensive footnotes to the poem are arranged to make a brilliant exposition of Newton's work and its place in history.
3. Perhaps the best work on the Declaration is still one of the older ones: Carl L. Becker, *The Declaration of Independence* (1922; reprint, New York: Vintage Books, 1943). For a different view, among many, see Gary Wills, *Inventing America: Jefferson's Declaration of Independence* (New York: Vintage Books, 1979).
4. For the most detailed review of the writings on this idea, see I. Bernard Cohen, *Science and the Founding Fathers* (New York: Norton, 1995).

5. Ibid.
6. See, for example, the discussion of nature in Charles A. Miller, *Jefferson and Nature* (Baltimore, MD: Johns Hopkins University Press, 1988).
7. William Blackstone, *Commentaries on the Laws of England,* facsimile edition (Oxford: Clarendon Press, 1983), book 1, pp. 38–54.
8. Ibid., p. 41.
9. Earnest Barker, "Introduction," in Otto Gierke, *Natural Law and the Theory of Society, 1500 to 1800* (Cambridge: Cambridge University Press, 1937).
10. Blackstone, *Commentaries,* book I, p. 120.
11. Ibid., p. 42.
12. Euripides, *Phoenissae,* ed. Donald J. Mastronarde (Cambridge: Cambridge University Press, 1984), p. 7.
13. Samuel Adams, *The Natural Rights of Colonists as Men: The Report of the Committee of Correspondence to the Boston Town Meeting, Nov. 20, 1772.* For this and other cases where Adams used the phrase, see Ira Stoll, *Samuel Adams, a Life* (New York: Free Press, 2008), pp. 51, 62.
14. Henry St. John Bolingbroke to Alexander Pope, in *The Works of the Late Right Honorable Henry St. John, Lord Viscount Bolingbroke* (Dublin, 1793), vol. 3, p. 343. The same sentence appears in his *Fragments or Minutes of Essays* (1740) and was copied by Jefferson into his Literary Commonplace Book, published as *Jefferson's Literary Commonplace Book,* ed. Douglas L. Wilson (Princeton, NJ: Princeton University Press, 1989), p. 40.
15. Alexander Pope, "Epistle to Dr. Arthunot" (1735), in *An Essay on Man* (London, 1735), epistle I, line 331.
16. TJ, "Opinion on the Treaties with France," *Boyd* 25, pp. 608–618.
17. TJ to Henry Lee, May 8, 1825, *LOC.*
18. TJ to Peter Carr, August 10, 1787, *Boyd* 12, p. 14.

EPILOGUE. Measuring the Shadow

1. Daniel J. Boorstin, *The Lost World of Thomas Jefferson* (Boston: Beacon Press, 1948), p. 244.
2. Charles R. Darwin, *The Descent of Man and Selection in Relation to Sex* (London: Murray, 1871).
3. Marie-Jean-Pierre Flourens, *Recherches Expérimentales sur les Propriétés et les Fonctions du Système Nerveux, dans les Animaux Vertebres* (Paris: Chez Crevot, 1824).
4. TJ to John Adams, January 8, 1825, *Adams-Jefferson,* pp. 605–606.

Index

Page numbers in italic *type indicate illustrations.*